PENGUIN BOOKS

THE HANDMADE HOUSE

Geraldine Bedell is a journalist for the *Observer*. She lives in London with her husband, Charles Leadbeater, and four children.

The Handmade House

A Love Story Set in Concrete

GERALDINE BEDELL

PENGUIN BOOKS

PENGUIN BOOKS

Published by the Penguin Group
Penguin Books Ltd, 80 Strand, London WC2R ORL, England
Penguin Group (USA) Inc., 375 Hudson Street, New York, New York 10014, USA
Penguin Group (Canada), 90 Eglinton Avenue East, Suite 700, Toronto, Ontario, Canada M4P 2Y3
(a division of Pearson Penguin Canada Inc.)
Penguin Ireland, 25 St Stephen's Green, Dublin 2, Ireland
(a division of Penguin Books Ltd)
Penguin Group (Australia), 250 Camberwell Road,
Camberwell, Victoria 3124, Australia (a division of Pearson Australia Group Pty Ltd)
Penguin Books India Pvt Ltd, 11 Community Centre,
Panchsheel Park, New Delhi – 110 017, India
Penguin Group (NZ), cnr Airborne and Rosedale Roads, Albany,
Auckland 1310, New Zealand (a division of Pearson New Zealand Ltd)
Penguin Books (South Africa) (Pty) Ltd, 24 Sturdee Avenue,
Rosebank, Johannesburg 2196, South Africa

Penguin Books Ltd, Registered Offices: 80 Strand, London WC2R ORL, England

www.penguin.com

First published by Viking 2005
Published in Penguin Books 2006
1

Set by Rowland Phototypesetting Ltd, Bury St Edmunds, Suffolk
Printed in England by Clays Ltd, St Ives plc

ISBN-13 978-0-141-01299-5
ISBN-10 0-141-01299-4

I

We didn't intend to build a house. We just wanted to move – somewhere a bit bigger, with more of a garden: commonplace enough ambitions, especially for families with young children, especially in Britain, where property is wealth, and so, in some vague and unexamined sense, identity. Not that we were thinking about anything like that.

It was December 1999 and I'd just had my fourth child. Ned was small and sucky, all mouth and blindly pawing hands, but he took up a ridiculous amount of space. He came with cot, high chair, bouncing deckchair for sitting on kitchen worktops, different bouncing contraption for dangling from doorways, Moses basket, pushchair, car seat, changing mat, monitors, sterilizing equipment, rattles, baby bath, specialist infant unguents . . . so much stuff that entire floors of department stores are given over to it, chains of shops have been established on the basis of it. Most of it was superfluous, but we had it anyway, and we were squashing it into a narrow four-storey house in East London which was already over-inhabited by our inconveniently sized family.

We are so inconveniently sized because we're what the sociologists like to call a 'reconstituted', or, more tweely, a 'blended' family. Reconstituted definitely does not sound like your authentic, ten-out-of-ten, top-of-the-range version; more a powdery, just-add-water-for-a-convincing-flavoured family. And blended is as bad: a euphemism for miscegenated, produce from around the world, thrown together in a factory staffed by the underpaid. Possibly I am over-reacting; but marrying twice was never a big ambition, and I am inclined to be a bit defensive about it. I am always resisting what I assume to be other people's conclusions that there is something illegitimate about us.

We are certainly untidily shaped. The two older children, from my marriage to a banker who now paints, were sixteen and twelve when Ned was born. Neither the banker nor I had ever, I suspect, been entirely sure why we were together. He occasionally expressed this openly, although the words didn't match his behaviour, which was intuitive and responsive. But perhaps he was more intuitive than he could cope with, and the words were not so much a joke as an attempted inoculation, fending off the whole truth with a taste of it.

Whatever, the consequence of my failure to take seriously the questions he raised was that when I had one child of seven, and another of three, I saw Charlie across a crowded room at a party. It can't have been that crowded, in fact, because I am rather short; it was probably more across a table. Charlie claims he fell in love with me at once, but he is lying: what he fell in love with was my dress. I'd bought it for my sister's wedding the year before – a dusky-pink silk jersey shift from Catherine Walker. Even though it had been in a sale, it was the most expensive dress I'd ever bought and it gave the wholly erroneous impression that I was classy.

I already knew who he was, vaguely, but I hadn't expected the jolt of recognition over the cheese sticks and Twiglets. He tried to get around the table to talk to me, but it was a long table: I turned and fled, found the host's elderly father, talked to him with concentrated effort for twenty minutes, and then left.

The relationship, Charlie's and mine, didn't start for nearly another year. But beginning in October 1991, we conducted a five-month affair which was perhaps mainly remarkable for the number of trees that had to be cut down to keep it going. During these months we wrote to each other with an energy that would have alarmed Barbara Cartland – long letters every day, sometimes two a day, pages and pages of handwriting on thick, creamy German paper. We were like an interminable eighteenth-century epistolary novel. Eventually, something had to give – quite apart from anything else, neither of us was getting any work done – and the only solution seemed to be to live together. That way, we figured, we wouldn't have to write every tiny thing down, and then analyse it and send a commentating letter back.

My husband was pretty helpful at this point. I had confessed to the affair about three months in, having always been a lousy liar, and initially he had scoffed that the relationship was 'just words' (hey, but a lot of them!). Some six weeks later, he acknowledged that the idea of our staying together was hopeless.

I continue to have mixed feeling about all this. A part of me would like to be able to write that I am ashamed, and I am, certainly, ashamed of marrying for the wrong reasons the first time, and sorry I had to kick around other people's investment (of which there proved to be more than I had realized) in my marriage. My children's lives were disrupted in horrible ways that are still unspooling. I don't want to pretend there's no sadness or guilt. But the main things I feel are relief that I found Charlie, a sneaking pride that I had the courage to live with him, and, finally a persistent and possibly irritating determination to prove that we are a successful family. (Charlie, incidentally, would regard this last as absurd: in his view we obviously are.)

For my generation, making a home is a more self-conscious activity than it was for our parents, for whom it was more or less a matter of buying a house and sticking your family in it. For me, it was even more than that: a test, and a proof. Creating a home together was a statement that it was OK, we were OK; it had been worth it.

By August 1991, Charlie had sold his workman's cottage near the Columbia Road flower market in Bethnal Green, and I'd sold my share in the four-storey terraced house in Islington to my former husband, who was living there with his new partner. Charlie and I bought another four-storey terraced house in the London Fields area of Hackney and, a year on, did what you were supposed to do to such buildings – restored the cornicing and the shutters, sanded the floorboards, reinstalled the marble fireplaces that had been ripped out during the period when the houses in the area had been little more than slums.

For as long as we had two children, it was fine. But then we had Harry in 1995 and Ned in late 1999. By this time, we were also both working from home, Charlie in our bedroom and I in the sitting room. I tripped over books with titles like *Liberals and Communitarians* and *Wellsprings of Knowledge* on my way from bathroom to hairdryer – interesting books, I am sure, but most saliently, for me, fat books. When Charlie sat on the sofa, he had to look at the pastel-grey, wire-sprouting back of my computer squatting toadlike on my desk, and at the box files in which I store my research for articles I've written. I can't remember ever having referred to any of this research subsequently, but I am too neurotic to throw any of it away.

In practice, we spent most of our time in the basement, which is where most converted four-storey mid-Victorian houses have their kitchens, and where we also had our dining table, television, most of the toys and a large, curved sofa, on which we could all just about fit. So, despite having a house that had three perfectly adequate floors above ground, we lived most of the time underground in a sort of troglodytic bunker, only dimly aware of the weather, or anything much except the passing legs of the neighbours.

The house was all stairs and, with four children and at least one untidy adult, stuff was always being moved on to the wrong floor (usually the basement, where there was no room for anything, because it was already full of people). I was for ever making little piles of toys, hairbrushes, books, ironing and dirty washing, about which I fondly hoped people would be responsible, collecting their possessions as they passed. In practice, everyone ignored them, perhaps imagining they were little votive piles for my amusement or spiritual enhancement, I don't know. It took me all of Saturday morning to put them back.

The original inhabitants of our house would hardly have been posh or rich. Not unlike us, then, except they could have expected to employ a servant to carry things around. I didn't want to be that servant. Nor, despite my family's fond suppositions, was I into purity through bending.

My skivvying predecessor wouldn't have had so many things to transport, anyway: the Victorians had fewer possessions, not least because you couldn't fit them in for the people. My grandmother, born at the beginning of the twentieth century, when our house would have been forty years old, grew up in a tiny factory cottage in the East End of London with three sisters and four brothers. My father, born in 1926, was the youngest of seven children brought up in another, very similar, East End cottage, and they managed to cram in two lady lodgers as well, in a room on the half-landing called the orff room.

At the start of the twenty-first century, by contrast, children expect their own bedrooms, music systems, miniature office space and walls to cover with Blu-Tacked pages torn from *Heat* magazine. When Henrietta, my eldest child, was sixteen, she kept complaining that she couldn't bring her friends home because there were too many little children in the house. My great-grandmother, the one who had four boys and four girls, actually gave birth to fifteen children. I was modest in my childbearing ambitions by comparison, but was regarded by my own daughter as more or less verminous.

We were using every inch of space, we believed, although it is conceivable we were overusing some inches and not using others very smartly. It was a philosopher called Daniel C. Dennett, however, who finally persuaded me that we should move, when I stubbed my toe on his book *Kinds of Minds*.

We had the house valued, pleasingly, although at this time it was virtually impossible to have a house in London valued unpleasingly. In 2000, property prices in the capital saw their highest ever recorded year-on-year increase, 23 per cent. In the previous decade – and we'd had the house for nearly eight years – prices had risen by 60 per cent. In the summer of the following year, which happened to be the Queen's Golden Jubilee, it was reported that over the fifty years of her reign, property would have been much the best investment any of her subjects could have made.

We still didn't really know how much more we could afford to spend on top of the capital we had accidentally accrued, because it was unclear how much Charlie was earning. Or even, in fact, what he was doing. In the first year after he quit his job as a newspaper executive, he wrote pamphlets with titles like *Civic Entrepreneurship* and *The Self-governing Society* (Henrietta and her brother Freddie referred to them as Charlie's leaflets). The second year, he did more consultancy, which was better paid, but the money was what I think bankers like to call lumpy, i.e. quite often non-existent. And if it was difficult even to explain what Charlie did – his mother asked my mother, who asked me, and I asked Charlie and, I expect, got a clear response, but not one that I could remember long enough to pass on to anyone else – it was a nightmare trying to gauge how much money he was liable to earn over the twenty-five-year span of a mortgage.

I visited estate agents, told them I thought we could afford £600,000, and explained our requirements: five bedrooms, one for each child and one for us (though six would be nice, a spare room always coming in useful), two studies, a large kitchen, which could possibly incorporate a dining room, and a sitting room. And a den to watch television in, and not on a busy road; and we needed a garden, preferably bigger and certainly nicer than the one we already had.

Some of the agents managed to suppress their smirks. 'But,' I said – this was my trump card – 'we're very happy to stay in Hackney.'

The London Borough of Hackney is widely believed by people who do not live there to be one of the most dangerous places in Britain. Or, in fact, the world. And they may have a point: it was reported in 2001 that Hackney was more dangerous than Soweto, measured by shootings. But we managed to remain oblivious to the Yardies, other than when they occasionally roared down the street in their souped-up black four-wheel-drives playing hip hop very loudly. There was, admittedly, the occasion when I arrived home on a sunny afternoon to find police swarming all over the garden, having just found £8 million of heroin, then the most valuable single haul in the UK, in the roofspace adjoining ours. The Turkish woman from next door, who spoke almost no English, was taken into custody and claimed in her defence that she'd been looking after the boxes for her boss and had no idea what was in them. She got off after her case collapsed because the police refused to disclose the source of their tip-off.

But even this impacted on us only to the extent that it created a mild ethical dilemma. Our neighbour was technically innocent, but she'd incontrovertibly had 8 million quid's worth of heroin in her loft. In such circumstances, was it appropriate to smile and say good morning, or to frown and look coldly disapproving? On a personal level, we'd always found the family cooperative. On the very rare occasions I'd asked her teenage sons to turn their music down, when, say, the baby was sleeping in the afternoon, they'd responded immediately. I guess if you have £8 million of heroin in your loft, you probably don't want to draw attention to yourselves.

Most of the estate agents dealing with our area were actually located in the next-door borough of Islington, to which Hackney owner-occupiers were deemed to aspire. Houses in Islington were very similar to our own, perhaps with the addition of another storey, rendering them even more inconvenient. The main difference was that they cost twice as much. Islington was already gentrified – as, in fact, so much the epitome of gentrification that it had become something of a joke, referred to in the papers as 'the trendy theme park of Cool Britannia' or 'the home of champagne socialism'. Not that Charlie and I would have minded any of this. We hadn't been at all gentrified as children. My family came from the East End and moved out along the Central Line with increasing affluence until, when I'd left for university, my parents finally made it over the borders into Essex. Charlie's family was Northern-respectable – Yorkshire-Methodist and Lancashire-Anglican stock, ambitious for the standing in the community that comes from plain living. We would have been quite grateful to be cool anything. But any house we might have wanted in Islington would have cost far more than we had to spend.

We did see an Edwardian house in Highbury that we liked. But having, as we thought, secured it at the asking price, we found ourselves in a bidding war with some people who'd put a note through the vendors' door and to whom, I suspect, they had been intending to sell the whole time. Our role, it seemed, had been mainly to jack up the price. Aggrieved, we refused to offer above a couple of thousand more, not realizing that in the fervid property jungle of the inner London suburbs in 2000, this sort of low behaviour was completely normal.

So we were happy to stay among the drug dealers in Hackney, where we tried the tactic with the notes ourselves, in a street where the houses were bigger than most. I trailed up and down with the pushchair and my handwritten requests on the German stationery. Not a single person rang up to ask how much we thought a 'generous offer' would be. Meanwhile, I had the sense that our combination of pickiness and poverty was starting to annoy the estate agents. There was a second competitive-bidding episode, for a white stuccoed house in a pretty street, where again we didn't offer enough and lost the house to some people who lived in our road in London Fields. (Nobody liked them.) But it was just as well, because we were supposed to be moving for more space and this house, as my sister brutally pointed out, was a cottage. House-buying is a potentially idiotic, brain-dislocating process, in which it is perfectly possible to get side-tracked by light through a south-facing window and streets laced with cherry blossom and overlook the fact that the house itself doesn't work.

So then we started wondering whether we should stay where we were, and somehow refurbish the shed at the bottom of our garden in a way that would get Charlie out of the bedroom. We already knew that it was unlikely that we'd be able to expand into the roof. We'd bought the house in 1992 hoping we could put in a loft extension, but Hackney Council's Planning Department had given us to understand that it wasn't even worth submitting a scheme. The house was on the end of a terrace and any addition would visibly wreck the roofline – and, since then, we'd become a conservation area. But maybe we could do something lower down. What we needed was someone who could see these things. What we needed was an architect.

We got two, Joyce Owens and Ferhan Azman. One was American, the other Turkish, they were both women, and beyond that we knew nothing about them except that they'd done our friend Hugo's kitchen and Hugo was almost certainly the most stylish person we knew. He was also a food writer, so presumably wouldn't entrust his kitchen to just anyone. (We later found out he'd chosen them mainly because he and Joyce had a mutual friend.)

Hugo and his wife Sue, who worked in the film industry, had children who were the same ages as our two younger ones, and we shared a school run. They lived a few streets away and we very much liked going round to their house because, apart from all the usual reasons – gossip – the food was simple but delectable, and the house itself was appealing: light, airy, invitingly spacious. There was a bench along the wall of the kitchen that more or less required you to sit down, and once you were sat, you didn't want to get up. The back of the house seemed open to the garden and, more abstractly, to possibilities (not least that if you hung around for long enough you might be offered some of Hugo's food).

This may be indicative of shallowness on my part, but when I think of people I know, when I summon up friends, they invariably arrive in my imagination against a backdrop of their houses. Without meaning to, I see them in kitchens, in front of fireplaces. I have one friend I always envisage against a background of lugubrious walls and dark carpets, of cabinets and fireplaces and not enough light, although I know that in reality he collects furniture and owns several pieces that are intensely covetable. (And he is, in all sorts of ways, a person who sees the grain of things, the detail, with powerful close-up vision, but as if he were somehow shortsighted when it comes to the whole. This means he is always interesting to talk to. But it also means he lives in a dark house.)

Another friend doesn't have a single stick of furniture that I could identify, were it to appear in front of me in a furniture identity parade. But I always think of him in a room on his ground floor, a room he has extended into the garden, where the experience of sinking on to his kitchen sofa and looking out at his plants is entirely consoling.

This can't be as eccentric as all that, or how else to explain the success of *Hello!*? and *OK!*, in which the main character is often the sofa? I am, for example, quite interested in the former *EastEnders* star Danniella Westbrook, because I've interviewed her and liked her, and because anyone who has destroyed their nose with cocaine is interesting. But I am *very* interested in Danniella Westbrook in her lovely home; it's fascinating to me that her small son has a fridge in his bedroom and that she (Danniella and I come, originally, from the same place) favours footballer's wife-meets-*Birds of a Feather* Essex-style. And this, of course, is snobbery, which goes some way towards explaining the whole thing. It no longer matters how you speak (my children, brought up comfortably in the ring-fenced stratum of Hackney middle-classness, speak variations of mockney; whereas I, brought up in the heartland of Estuary English, sound rather posh, although I increasingly try not to, because, like bad teeth, it dates one so. No one under twenty-five is posh.) What your parents did doesn't matter any more, either; these days you are supposed to be able to make your own life. But where you choose to live and how you opt to decorate your house is a symbol of – not class, exactly, but something like it; what matters to you.

So, assuming that Hugo knew what he was doing, we invited Joyce and Ferhan to come round, explained the problem of having too many children, and let them wander about. Joyce, the American one, was plump and funny, easy-going and open; Ferhan was small and dark and fierce, with a spitfire delivery. I trailed behind them, registering properly for the first time in a while the not-very-good Persian carpets, the chaos of furniture, some of it Arts and Crafts, some vaguely Far Eastern, picked up when I lived abroad, some from Heal's and Habitat, in mahogany, old oak, limed oak, or, occasionally, cherry; the hectically patterned red and gold sofa (much later, Joyce would scathingly dismiss a much less plump sofa to which I'd taken a fancy as 'overstuffed'), the other sofa, striped in a manner reminiscent of Edwardian seasides, jolly awnings and cheery deckchairs; the boxes of toys in plain view because there was no cupboard space for them; the wardrobe doors with bits of dowelling tacked on to them to resemble, in theory, panelling; the paint-effected bathroom walls into which we'd been talked in a moment of exhaustion by a decorator who'd just caught up with the rag-rolling and sponging fads of the 1980s, sadly only a decade too late. Clearly, we hadn't got a clue. 'I like the lamp,' Ferhan said eventually.

That was it: the lamp was OK. The rest of it was a mess of stuff dragged in from my former life, Charlie's previous bachelor existence or, more rarely but no less eclectically, bought together. We had, it occurred to me, an awful lot of stuff, and none of it went with any of the rest. It was all jostling for attention, cluttering up the view and scrambling our brains. Perhaps, I thought hopefully, when Henrietta had said she couldn't bring her friends round because there were too many little children, what she *really* meant was that there were too many patterns and clashing colours.

Joyce and Ferhan saw what we should do immediately: put the offices underneath the garden. As they explained it, you'd come out of the back of the semi-basement into a courtyard, and instead of coming face to face with a brick wall and steep steps up to the pavement-level garden, you'd see glass doors leading to our offices (I could have an office too!) which would also be lit from slots above, in the garden. Essentially, they were proposing digging the garden down to basement level, building a couple of offices, and putting the plants back on top.

You could tell it would look fabulous. I could envisage it immediately, all clean lines and cool spaces, how restful it would be and how dedicated to work. This, clearly, was what you employed architects for: they had an ability to reconfigure space in their heads, not to get stuck on what is, or see earth as immoveable, or live with those inconveniences around which we had bent ourselves until we were ossified in contortion.

There was only one problem with Joyce and Ferhan's inspired scheme: it would make the rest of the house look rubbish. You wouldn't want to emerge from those sleekly beautiful offices into our battered kitchen and higgledy-piggledy dining room/

den/playroom. Joyce and Ferhan thought we needed more glass at the back of the basement – all the way across instead of French windows – and then you'd want to put the dining table in front of it, which would mean moving the kitchen to the front . . . It was like aesthetic skittles. And it was going to cost a lot of money.

I loved that house, for all sorts of reasons. It was the first house that Charlie and I had owned together. Henrietta and Freddie had effectively grown up there. Their father had married a politician, and they'd divided their time between the constituency and various houses in London. The children had shuttled about with them, exhilarated by the changes of scene, but coming to rely more and more on the house in Hackney as their fixed point. Their best friends, a brother and sister, the girl Henrietta's age and the boy Freddie's, lived across the road.

Charlie and I had had two more children there; I'd been in labour on sunny mornings with the light streaming through the long, east-facing bedroom windows and the cat at attention beside me, concerned, on the bed. My sister lived across the road, following an established East End tradition of families all living in the same street and for ever being in and out of one another's houses (to which Elaine and I weren't in reality in the least attached, having grown up in the suburbs). Harry and his cousin Flo had been born only ten days apart, had grown up together and were still each other's default playmates. But when, as tended to happen more and more, they got fed up with each other, it was easy to come home in a huff.

All the same, I wasn't sure it was worth spending the sort of money that Joyce and Ferhan's scheme would entail on this particular house. In Britain, or at least the owner-occupier part of it, houses have a dual, sometimes contradictory function. On the one hand, they are the investment in which the majority of family capital is tied up: a store of wealth and a means of creating it; a status symbol, to be traded up. On the other, they represent stability and security, places of retreat in an increasingly hostile and precarious world. (Sales of cushion covers and scented candles reportedly quadrupled in America after September 11th.) And so maybe it's not entirely fanciful to imagine that the vogue for house and garden television makeovers is at least partly a response to the uncertainties brought on by globalization, the feeling that decisions are made far away, that El Nino-style forces are at work on our lives. The only place we're in control is at home. But it's a lot to ask of houses, to be a bulwark against insecurity, if we're always looking to trade them up, move on, acquire a shinier new one.

In the end it was the cars that decided me. It would have been sensible to spend the money if we'd been prepared to stay in the house for another decade or more, to decide that this really was our home, was where we belonged, how we wished, finally, to be defined.

But there was too much wrong with it. The house was on a corner. The street we were in was quiet; the one that ran along the side of the house and garden, to which in many ways we were more exposed, was getting busier all the time. Officially, this side road was classified by the council as residential, but it was also an emergency services route, which posed problems whenever traffic calming measures were proposed (especially the one we favoured, of closing the road altogether). One of the councillors who took a particular interest in traffic lived a few blocks up, in the street that was classified as the main east–west route through the district. But this road had had so many traffic-calming measures imposed on it, including being mysteriously closed for months, that our road had taken the weight of the extra traffic and never lost it. And there was, unfortunately, no shortage of extra traffic. By the time we were debating whether to move or stay, the M11 extension had just opened, meaning that more and more traffic was pouring into London from the east all the time. Hackney Council had published no strategy to deal with the extra vehicles heading through towards the City and it gradually became clear that the Town Hall didn't have one. Or not officially, anyway. Unofficially, it seemed to be to send the traffic down our road.

By that stage, we had been campaigning about the traffic for several years (some people vigorously, but I'd tried to turn up) and we were getting pretty despondent. The council had put humps in the road, but they were the kind that you can get your wheels either side of and, since all the cars tended to be going in one direction – into town in the morning, out in the evening – this had zero effect on slowing them down. We couldn't interest the neighbours in the roads running parallel in our campaign because they worried that any restriction for us would just shift the traffic on to them.

I started counting cars. I'd stand at the bedroom window at rush hour, when I was supposed to be getting ready to take the children to school, muttering madly, 'That's one every five seconds.' I'd prowl up and down the long sitting room in the middle of the day, trying to find a minute when there was not some vehicle passing. I made notes and showed them to Charlie, who looked concerned (for me, not about the traffic). I started shouting at drivers (who couldn't hear, since they were in cars and I was in the house. But the point was, I could hear *them*). The state of the traffic became a persistent topic of conversation for me. I faced the possibility of soon having no friends because of being so boring. I considered a future of summer afternoons in the garden frothing at the mouth because of White Van Man rattling past my clematis. The more I thought about staying where we were, the closer I got to being hospitalized.

We went back to looking for houses.

And it was depressing. Since we'd come close to buying one Edwardian red-brick house, all the estate agents (we were, by now, on everyone's books) assumed that we would buy another – any other – and never mind that the one they were showing us had been done up in a way that was both expensive and shoddy by some developer, or that some other one was dark and had a tiny concrete yard instead of a large, sunny garden and was inhabited by funereal Italians with antimacassars and packed suitcases in the parlour, presumably ready to flee somewhere more cheerful as soon as someone would give them some money.

We were being difficult, I knew. The more we looked, the pickier we got. Even I was starting not to like us. So when Sue Reynolds from Currells called that Tuesday morning in May, she must, I think, have had another buyer in mind. After all, we had a great track record of coming second. We were probably becoming quite a useful fixture on the Islington estate agents' scene: the people who'll push the price up.

'How,' Sue asked, 'would you like to buy a piece of land?'

At any other point of my life, I'd have said, 'No thanks, what would you do with that then?' I was not one of those people who trail around looking at old garages, wondering if they can be knocked down and replaced with a fabulous cutting-edge contemporary dwelling of their own design, for the simple reason that I could no more envisage a fabulous cutting-edge design than I could revise Einstein's theory of relativity. I have no visual imagination. It is questionable, in fact, whether I have any spatial awareness; I quite often bump into things. I became a journalist because it is one of only a very few careers open to someone who can't do either numbers or pictures. As a child, I learned to read early mainly because the illustrations didn't make that much sense to me; it was the only chance I had of working out what was going on. Even now, when my younger children ask questions about pictures in their reading books, I am bemused. Quite often I haven't noticed that there are any.

So I would have assumed, if I'd ever given it a moment's thought, that building a house was something done by people with an artistic side. I didn't have so much as an artistic pockmark.

But this was not any other point of my life. It was the point at which I had discovered Kevin, which had changed a lot. Kevin (as I liked, familiarly, to think of him) was the tall, handsome, dark-voiced presenter of *Grand Designs*, a television programme that had first aired the preceding autumn, and to which I had been glued, in a heavily pregnant, unable-to-get-off-the-sofa kind of way. *Grand Designs* followed the thrills and vicissitudes (mainly the vicissitudes, although I didn't register this at the time) of people as they built their own homes: barns in Oxfordshire, straw bale houses in North London, or communes of eco-houses with composting toilets on the South Coast.

Kevin was so athletically enthusiastic, so sexily knowledgeable – outdoorsy but intellectual, sympathetic but sceptical, handsome but clever, muscularly articulate – that he made building your own house seem not only possible but really, really attractive. It is possible that I assumed that if you built your own house, you automatically got Kevin. He'd be there in my kitchen eating Hob Nobs, dropping the crumbs on my plans, admiring my taste in kitchen worktops and discussing modernism with me. And then he'd be up and away, swinging around in my rafters in his hard hat and speaking to camera like an architectural Tarzan.

Nearly as importantly, the self-builders featured on *Grand Designs* all seemed quite ordinary. At least, the people who were building the house out of straw bales were a bit weird and the composting toilets lot clearly had their own way of looking at the world, but the point was that they weren't millionaires. One of the few practical pieces of information I had managed to glean from *Grand Designs* was that you could build a house from scratch for less money than it would cost to buy one. Building was cheaper than refurbishing because you didn't have to deal with the complications created by the pre-existing structure and you don't pay VAT on new build.

I called Joyce and asked her to come and see the land with me the following morning.

*

The plot was really just the scrubby end of a garden. We approached it from the house to the north, a wide 1950s villa situated in the road behind, which was privately owned, for some obscure historical reason, and policed by the owner-residents with ferocity: if you parked here for more than five minutes, you were liable to be clamped. We parked anyway, and went in through the side gate.

Most of the large garden belonged to the house. There was a wide terrace, with a magnolia tree, borders sprinkled with muscari and irises, a top-of-the-range children's climbing frame, and, at the bottom, a patch of nettles and scrub.

We stood well back from this bit of scrub, on the tended lawn, not wanting to venture on to the uneven ground with its layering of weeds, ferns, brambles and whatever might be scurrying or slithering through them. There was a straggly, self-seeded fruit tree, clover, dandelions and a rusty garden roller. The sun was shining and the air was warm enough to wear a T-shirt; it was one of those days when you realize with relief that summer has come. I stood there in the sunshine listening to birdsong and feeling obscurely happy, seized by certainty.

The land came with planning permission for two houses – 'developer's houses', was how everyone referred to them, dismissively – and Sue Reynolds's understanding was that Islington Planning Department and the local residents would look more favourably on a single family house.

There were mysteries. Why was Sue Reynolds offering this to us? She must have had plenty of clients with more money. We had shown ourselves incapable of following through with any purchase, or even being very sure of what we wanted. There are people who plan to build their own houses years before they even have a sniff of a site; who tour the country at weekends, peering keenly out of their cars, looking for a place to put their dream. We just wanted a house to contain our children and their vast amounts of stuff.

The people who are after sites can rarely find them, because developers get there first. In this case, the couple who were selling the house were property developers themselves, so maybe they knew that the price they wanted to charge for the land (£500,000) was too high to make commercial development worthwhile. So they were selling through an estate agent in the hope of luring in some clueless novices. Maybe we looked clueless. We only had an indistinct sense of how much we could afford; possibly they thought they could persuade us to pay too much.

I honestly don't know; but I did realize that this was a chance like no other. The land was seconds away from Highbury Fields, one of the few green spaces in Islington, and the nicest, surrounded as it is by fine Georgian terraces. It was minutes' walk from the Tube, and bus routes ran along the top of the lane. Yet you could stand in the garden and hear only birds.

The second mystery was the history of the land. From what Sue Reynolds told me (I possibly wasn't listening all that carefully) I somehow formed the impression that it had only been incorporated into the garden of the house in the last couple of years, subsequent to planning permission having been granted in 1999. The vendor seemed to reinforce this – implying, when I met her a few weeks later, that she and her husband had acquired the land with the idea of building a swimming pool, but that now they were moving to send their daughter to a Steiner school in Sussex.

But it's perfectly possible I was only half-hearing, or hearing what I wanted to hear. I also managed to absorb the information that the house that was on the site originally had been bombed in the war, and nothing had been built here since. This made sense, because it looked very like the bombsites I remembered all over the East End from my childhood, sprouting buddleia, looking as if they were struggling to recover their lost identity but didn't know which way to turn – towards brick and buildings, or back to nature.

'Would it work?' I asked Joyce. 'Could you build a house here? Big enough? What d'you think?'

What Joyce privately thought, as she later told me, was that we would never build a house here, because we hadn't seen the lane yet. It was all very well to come at the site through this splendid villa with its big south-facing garden with fancy climbing frame; actually, the approach would be down a narrow gravelled lane studded with potholes and bordered by factories on one side and wire netting on the other.

'Yes,' she said, 'sure' – because how often does an architect in London get the opportunity to build a new house? What was she going to say? 'Who wants this kind of work anyway?' But I was surprised, because I looked at the site and had absolutely no idea whether a house would fit on to it or not. How much space did houses take up?

Joyce suggested that we should come at the site from the other direction, so we got back in the car and bumped down the lane, over the rough ground. To the right was a former factory that had been converted into workshops and light industrial spaces. A couple of these had back entrances on to the lane, for collection and delivery; I was conscious that it was fortunate there weren't any lorries collecting/delivering right now because if you met one, the only way out would be backwards, reversing through the potholes. To the left was a modern brick-built house, which had been constructed in 1993 and looked as though it really belonged to Surrey; and beyond our plot, a two-storey Victorian building known as The Glassworks, formerly a small factory, but now empty, after which the lane petered out in a dead end. Opposite the Surrey house was another empty site, much longer and narrower. This had recently been acquired by an architect who had designed, and got planning permission for, a modern house.

The lane was a nightmare: narrow, with no turning points. It was lethal to tyres, and it didn't feel remotely like the wide and elegant avenue behind; it felt like the side road to a factory. 'I like it; it's edgy,' I told Joyce.

Sue Reynolds said the address would be 1 Ivy Grove Lane. That decided it. It was too nice an address to pass up. 'I want it,' I said recklessly, oblivious to the first rule of house buying, i.e., don't look too keen. Charlie hadn't even seen it. In the sunshine, squinting at the rubble and ferns, I offered the asking price.

2

December 1999 was significant for something apart from the birth of Ned, though its impact was obscured for a bit by sleeplessness and the hazy cocoon of the babymoon. Earlier that year, Charlie had published a book called *Living on Thin Air*, which attempted to explain the growth of the knowledge economy (fewer people working in manufacturing; more and more of us spending our time juggling bits of information). The book was pretty philosophical in spirit, and about all sorts of things, including the importance of creating equitable social structures to cope with a new economic order. But its appearance coincided with the dotcoms boom, and it was taken up by some internet entrepreneurs as a sort of bible for the bubble.

A firm of venture capitalists specializing in investments in new technology approached Charlie to work with them – which was a brave move on their part, since he knew absolutely nothing about either investments or technology.

As far as the former were concerned, the tattered remnants of his former communism clung on as an almost principled lack of interest in ways of making money from money (as opposed to from work, to which he wasn't opposed at all). We had a house, of course, carelessly accumulating capital, but we needed that to live in, so it didn't count. Certainly, he wouldn't have contemplated applying for any of the handouts disguised as share issues created by Thatcherite privatizations. And although he has worked for the *Financial Times* – where he'd been industrial editor, head of the Tokyo Bureau and features editor – and now writes books that are perceived as being about business, he is frankly uninterested in capitalism as a personal venture and probably, at some level, finds the notion obscurely distasteful. For our first wedding anniversary, I mischievously, and I thought rather subversively, bought him a B-class share in Warren Buffett's investment company, mainly because I'd recently been to interview Buffett in Omaha and thought he was clever. Charlie was polite about this present, but that's all. I think I can honestly say that he has never shown the slightest interest in the performance of this share and always throws away the letters that come about it unopened.

But the fact that Atlas Venture thought he was insightful and brilliant counted in their favour, and Charlie negotiated a one-day-a-week contract, which left him the rest of the week to do other things.

By May 2000, when I first took Charlie to see the land, the day after I'd seen it myself, the benefits of this slug of money from Atlas hitting the joint account every month were beginning to tell. The telling took the form of a general sense of there being some slack; of not waking up in the night rearranging figures in our heads in an effort to make them come out less frighteningly. We were already better off than we'd been when we started house-hunting six months earlier, and that gave us the confidence to believe that we could borrow enough to build, even after paying as much for the land as we'd originally intended to pay for a house that was already standing. Quite where the money would come from wasn't clear: someone, presumably, would lend it to us. I had a recollection that Kevin had confidently asserted in the course of one of his programmes that a main contractor's quote of £60 a square metre was reasonable. This meant that a 2,500-square-metre house (I had no idea how big this was, but it sounded pretty large) would cost only £150,000 to build. In some parts of London, people think that's cheap for a conservatory.

The Elektra House in Stepney, which was then in the process of construction and has subsequently been shortlisted for all sorts of architecture prizes, was completed for a grand total of £70,000; the brief stipulated that it should be no more expensive, metre for metre, than a standard Barratt home. More than one couple on *Grand Designs* had built their own houses because they couldn't afford any that were already standing.

Plus – the reasons to pursue this were positively piling up – there was the business of not having to pay VAT on new build. So we ignored the fact that the only people we'd ever met who actually had built their own house were richer than us: richer, in fact, than we could ever imagine being. Their – presumably very expensive – house was perched on a cliff overlooking the sea and featured a glass dining-room wall that dropped down into the cliff, so that you could eat inside while feeling you were outside, floating over the water. We'd visited once or twice, when we were staying in bed and breakfasts nearby, and I'd driven away awed by the idea of being able to build something so fabulous, so individual and apparently enduring. Now I casually elided that house with the £100,000 houses off *Grand Designs* and assumed that we could have something similar.

Immediately Charlie had seen the land and agreed with me that we should go for it, we approached Lloyds Bank. This seemed the obvious thing to do: they'd arranged our current mortgage and Charlie had banked there since he was at university. We thought, naively, that this might count for something; but if there ever was a time when bank managers knew their clients, when they made decisions influenced by the duration and stability of relationships, that time is long gone. For most people, the non-fantastically rich, banking decisions are a matter of rules, not individual judgements. If you can get past the automated telephone systems with their multiple choices designed to encourage you to give up, the human beings you encounter are all reading from scripts. Pathetically grateful to have got through to a person, you accept that it would be asking too much to expect them to know anything about you.

The rules according to which Lloyds ran their business didn't permit lending money to individuals for the purpose of buying land. The best they could offer us was an extension of the mortgage in Hackney, so that we could withdraw some capital.

A friend of a friend suggested that we should approach a mortgage broker, who might have more options. She recommended someone she knew. I called this man, Peter, and he listened to my gabbled account of how much we earned and what we wanted to do, said it sounded reasonable enough, and promised to talk to his contacts. In the meantime, he urged us to prepare our accounts, because we'd need to prove we were really earning the money that we claimed. I am a person who instinctively hides bills underneath piles of newspapers in the hope someone else will accidentally throw them away and I can forget they ever arrived. It is almost a point of principle not to send in my accounts until my accountant has threatened to fire me. But on this occasion, I took the Nurofen upfront and got down on the floor with the till receipts.

By now, though, someone else had seen the land and offered even more than the already hefty asking price: £525,000. We raised our bid to £550,000 – still without actually having any money – but emphasized that this was absolutely as high as we could go. Then we were told there would be sealed bids, due by noon the following Tuesday.

But hey, conversations with Peter were always upbeat. At different hours on different succeeding days, he mentioned (always with the same levels of excitement) negotiations with Northern Rock, the Halifax and Barclays. Then Barclays got mentioned again. And again and again. He was bowling us along with his enthusiasm for the deal, glissading over the details. We consented to be bowled, aware that we were all in the business of boosting each other's confidence. This was a high-wire act and looking down could be dangerous. Anyway, no need to worry, because the deal Peter had cooked up was great, great, and only dependent on the production of our accounts and the planning permission.

The accounts. Still suffering from a headache, I drove through the heavy traffic that was clogging the early summer air with pollution, carrying boxes of till receipts, remittance advices and bank statements – now roughly sorted and vaguely understood – to our accountant in Finchley. I staggered into her office through the fumes that were masking the other summer smells of mown grass and blown roses, dumped the unwieldy pile on her desk and pleaded with her to make sense of it in three days.

That, then, left only the planning permission. Now and again, I'd mumble incoherently to Charlie something along the lines of did he not think that it might be a problem, this planning permission thing? He had now largely taken over the Peter-discussions, because he could remember numbers whereas I am a kind of numbers-dyslexic, and he swatted these anxieties away. We *had* planning permission for the two developer's houses. We knew Islington Council and all the neighbours who had objected last time would look more favourably on a single family house. The land was good for building, therefore was valuable. It could hardly matter to the lender what shape our building would eventually take.

That weekend, we went up to Yorkshire on a long-planned trip to see Charlie's parents. They live in the Dales, in a small four-bedroom house in a new cul-de-sac development. The houses are build along traditional lines, i.e. with small windows to resist pre-central heating-era weather, so that, despite having some of the most magnificent views in Britain, you can't see them. Since there's no longer much work in the Dales, the villages are now mostly inhabited by old people. Occasionally they back their cars out of their garages a bit too hard across the narrow cul-de-sacs, accidentally into one another's front rooms.

We sat in Charlie's parents' front room, watching out for senior citizen motorists hitting the wrong pedals on their automatics, and Charlie said he thought we should decide whether we really wanted to do this, to build a house on this bit of land. And the more we thought about it, the more convinced we were that we shouldn't pass up the opportunity to design a house around our own idiosyncrasies, rather than trying to cram ourselves into some Victorian or Edwardian notion of what a family should be. A house that we designed for ourselves would have no wrong-sized bathrooms, no alcoves you couldn't see the point of; all the space would work for us instead of competing with us, snagging us, leaving us feeling frustrated.

Building, rather than buying, would solve all sorts of problems. Henrietta wouldn't have to occupy a bedroom directly above ours, meaning that every time she played a CD, or even padded about after my bedtime (10.30 p.m., but look, I had a baby), she woke me up. Plus, Charlie and I could both have offices. No more working in the sitting room. All the children could have their own bedrooms. We could have walls of glass and, if they didn't quite drop into the ground, they'd still be lovely.

In that case, Charlie said, it was clear that we should pay as much as we could afford. The question was: how much *was* that? Neither of us was clear. So we sort of thought of a number and doubled it.

Or Charlie did. I never would have dared. But he said he thought we should bid £650,000. This was, frankly, a ridiculous sum of money. Even given our previous sorry history of competitive tendering, offering £150,000 over the asking price was an extreme gesture. Apart from a tiny fraction that we'd managed to save, mostly since Charlie's independent career had started to pay, it would all have to be done on borrowed money. (We both conveniently overlooked the fact that, at that stage, nobody had actually come forward offering to lend us any money at all.)

Charlie has said subsequently that he recognizes now that he was influenced by the six months he'd already spent at Atlas Venture, where people were always doing deals with absurdly large sums of money, as if millions were a minor thing, not scary and likely to end with your serving a prison sentence. He'd become infected with a certain kind of pig-headed entrepreneurialism.

This is true, I'm sure; but it's also true that this flash of bravado, this absolute self-belief, was entirely in character. Charlie is in many ways the most unassuming person, modest and restrained; I've never seen him boastful or bombastic or belligerent. His brilliance is combined with understatement and grace; he does not, I think, consider being brilliant very important. All the same, he's capable of fixing on something and believing in himself to a degree that is close to madness. He did this when he met me. I remember us sitting in the sunshine in Gray's Inn, in the early spring of 1992, a few months after we'd met, and Charlie telling me that he could see in my face what I must have looked like as a little girl, and how I would appear as an old woman; it was all there, in a flash, illuminated by the early sunshine, which was struggling to be fierce: it was there in my freckles and lines, what I'd been and might be. I knew he was telling me that he loved me, not in spite of my having married someone else and already having two children, but because of it, because that was who I was. And he was also telling me that he would go on loving me when I was old. He could see me as a lined and tired person, not vigorous, and he was promising me that he would feel the same then, too. It was impossible to resist (even if I'd wanted to) such clarity and certainty.

And this was a bit similar. So, the following Tuesday morning, I took an envelope into the estate agent's in Islington promising to pay £150,000 more than the asking price for a patch of nettles.

All this – seeing the land, getting turned down by Lloyds, approaching the mortgage company, asking our solicitor to start a search, preparing our papers so that accounts could be drawn up, and deciding on the size of our tender – took only a week. Charlie, meanwhile, was in the middle of preparations for a trip to America, where *Living on Thin Air* was being launched at the Chicago Book Fair, and was spending most of his time setting up newspaper and radio interviews in the US. I was dealing with four children, the oldest of whom had just transferred schools for A-levels and started what would become a three-year relationship with her first serious boyfriend, while the youngest was just sitting up and gurgling. I also had a full-time job.

We were pretty busy – and that, presumably, is why neither of us clarified exactly what sort of planning permission we would need to secure the loan. We found out soon enough. An hour and a half after I'd delivered the tender, Charlie called me: I was at home, he was in the departures lounge at Heathrow, waiting to board his flight to Chicago. He'd just finished a conversation with the mortgage broker, during which Peter had finally come clean, or we'd finally faced up to the truth; whatever, even with all the can-do brio that had been in the air the past week, we couldn't go on kidding ourselves any longer. Barclays needed planning permission for the house we wanted to build. Having planning permission for two other houses that nobody now had any intention of building wasn't going to get us any money at all.

Obviously, we were nowhere near having planning permission. We hadn't even briefed the architects. We hadn't even *appointed* them. There were no designs; no one knew how big the house would have to be, or how many rooms it would have, or what materials it would be made of, or what style it should be. It was less than a figment of our imagination.

'That means,' I said, trying to get it straight in my head, 'that we've just offered to buy the land with £650,000 that we don't have?'

'Right. At the moment we don't have it.'

'So where are we going to get it from? Peter's already spoken to those other companies, the Northern Rock/Halifax people.'

'He's talking about a bridging loan, to take us up to planning permission.'

'Could we manage that?'

'I don't know. I've got to go, my flight's being called.'

'Oh well, maybe we'll be outbid.'

An hour later, the estate agent called to say our bid had come out top – 'Not by much,' she added consolingly, but then presumably that's what they always say: if you realized you were paying £100,000 over the market rate you might back down. We were the proud owners of a patch of overgrown weeds and an ash tree with a preservation order on it – or we could be, if only we could magic up the money to pay for it.

In fact, of course, we were no nearer to owning it than we had been before we realized it existed.

'Now, did we explain to you,' the estate agent inquired smoothly, 'that it's a condition of the sale that you exchange and complete on the same day? And the vendors want the whole deal – the house behind, and the land – to go through at the same time? And they want it done within a fortnight, or the deal's off?'

'That's, like, a fortnight from now?'

'Exactly. A fortnight from now.'

No, they had not explained this. This was actually the first we'd heard of it. This development only confirmed my paranoid suspicion that they'd wanted someone else to have it and now, unable to refuse our massive, oversized bid, they were laying landmines in our path. Not that we needed landmines: we were perfectly capable of blowing ourselves up. When Charlie next called, now from Chicago, he revealed that Peter's bridging loan would cost us £6,000 a month. That was £6,000 a month that wouldn't actually be spent on anything – not repaying the capital or building a house, just the privilege of having the loan.

Not only had the existing planning permission not helped; it had caused us problems. There was planning consent for two developer's houses; therefore we must be developers. The only sort of bridging loan we could have was a commercial one. Even Peter dropped his gung-ho act to acknowledge that, actually, this wasn't the sort of loan that a couple with no real capital, four kids and erratic incomes should really be taking on. His, and our, high-wire act was over. So we might as well have had six months to exchange and complete. We'd made the bid without the money to pay for it, and he'd exhausted his options. It only heightened the absurdity that in reality we had a mere two weeks, for most of which Charlie would be in Chicago.

Over the next few days, when Charlie wasn't out promoting his book, he was sitting in his hotel room adding up figures – our savings, the mortgage advance from Lloyds – in a grim effort to make them somehow come out differently. He admitted subsequently that he was a bit blasé during this period, but he says he never doubted that we'd find the money. He was faintly surprised when I said to him down the phone line from London that I admired his determination but I just didn't think we were going to be able to do this. (Translated out of loving-married-people speak this reads as something like, 'Why are you being such a time-wasting prat?')

In a sense, of course, he was right: we'd already accomplished the most extraordinary and difficult thing – we'd found a plot, exactly where we'd have chosen to live, and secured it. Compared to that, getting a loan was a commonplace, everyday experience. In another sense, of course, the land had fallen into our laps through no effort of our own and was not an accomplishment at all. And the everyday, commonplace experience of getting a loan was not coming to pass for the very good reasons that we weren't a good risk, the land wasn't a good risk, and we might well be deluded about the prospects of planning permission.

All Peter could now suggest was that we might be able to raise a bridging loan through Lloyds, on the basis of a mortgage offer from Barclays subject to planning permission. We could certainly get a letter of intent from Barclays saying that they were prepared to lend us the money to buy the land eventually, once we had a design for a house and permission to put it there; that they'd gone through the full costings and judged us a decent risk – which, he said, might be a 'comfort' to Lloyds.

But it wasn't. Lloyds weren't comforted, weren't interested in the least. We were stuck.

Except. Except, except . . .

My sister Elaine had no mortgage. This enviable state of affairs had arisen because, in the early 1990s, she'd been involved in setting up and running a television production company which was subsequently sold, and she and her husband Clive had invested the proceeds in owning their house outright. And with typical generosity, which all the same stunned even me, since I am inclined to have a haughty older-sister-taking-for-granted approach to her, they offered to make the deeds over to us, so that we could borrow against it.

She suggested I call our bank manager. We have the same one, because we both bank at the branch of Barclays where our father's (and later, when he died, our mother's) office audio equipment business had had its account – always, this being my parents, respectably in the black. I'd met Steve Symonds only once, briefly – we had nothing to discuss, since I had no money – but Elaine knew him reasonably well.

So I called him. It was necessary for me to do this rather than Charlie, a) because it was my bank and b) because Charlie was in Chicago, but if I could have got out of it, I would have done so. As I have a lot of trouble with numbers, I try to avoid situations where this might be exposed. It's not that I'm incapable, exactly, just a bit simple. I can see the pleasure in adding up columns of figures to make them agree. (My mother was a comptometer operator as a young woman and, had she been born thirty years later, would probably have been an accountant or computer programmer; for her, a column of figures tallying with another column of figures is a thing of beauty. Literally, in fact, as she has scrupulously elegant writing, while even my written figures seem to struggle to evade my grip.) I can read a page of text with one figure on it and remember everything about it except the number. I'll be aware that there was one, just not how many noughts it had on the end. It is a humiliating condition; the only times I've ever come close to making mistakes in journalism have been over this difficulty with noughts. Numbers get into my head and start throwing a fog around, and that makes me anxious and distracted, so I don't concentrate properly and make matters worse. The prospect of having to make a pitch on the basis of numbers was panic-inducing. But there was no one else to do it. And I was guiltily aware that if I hadn't been so useless and terrified, I would have done it before.

Steve Symonds was an ordinary bloke, as well as a genius. He'd worked for Barclays for twenty years, since leaving school, and he lived in Chislehurst. These days he worked mainly in the high end of the business, dealing with individuals with incomes over a certain amount – I never found out how much, because I started gabbling as soon as he answered the phone, before he could kindly explain that I ought to talk to someone else. But Steve listened. I suppose, if I am honest, his enthusiasm began when I started talking about Charlie, or more specifically, about Atlas. But the first thing he actually said was, 'I don't think we need to involve your sister in this.' The whole complicated thing with the deeds wouldn't be necessary.

The second was, 'You'll need to provide heavy-duty paperwork, because I don't know Charlie'; and the third, 'You thought you had financing in place, but it's fallen through so now you're back to basics' (thus neatly demolishing my confected justification for bidding for the land on the basis of fictitious money). He saw through us.

'If I did this,' he concluded, 'I'd want to do the whole proposition. You *could* finance your build with Lloyds, but doing the back end as well might make it easier for me to get the bridging finance.'

It was probably around this point that I realized he wasn't just being polite; he was really interested. He could see that it might work, might even make him some money; he got it.

Already, Steve was thinking about what we'd need: a valuation on the land – he was somewhat surprised we hadn't seen one before making our offer – and a projected valuation 'on a build-out basis', i.e. one that would tell us how much the finished house was likely to be worth. And we should bring along details of our existing life cover, pensions, anything that would help create a complete picture of our finances. We were, he said, only going to get one go.

He suggested we come and see him as soon as Charlie got back from Chicago, bringing evidence of our income, proof of any more due in the future, our accounts and Charlie's passport and driving licence to prove he was who he said he was. And then we should put together something about what had persuaded us to buy the land at this price. When I stuttered out my explanation of this he said, 'Yeah, the more we talk, the trickier it's becoming.'

Steve Symonds was recognizable to me; he reminded me of a lot of the boys I'd gone to school with in the East London suburbs – clever, unpretentiously ambitious. I always felt different from those boys when I was an adolescent, although I didn't know why. Perhaps it was simply that my family was first generation middle class and I had an almost immigrant sense of the possibilities of displacement.

Friends Reunited means I can look up those boys now and

find out, roughly, what they're doing. Many of them still live locally, or in places remarkably similar to where we grew up. And I still don't really understand why this makes me despondent. There were plenty of things about the suburbs, those rows of 1930s and 1940s terraces with bay windows and their dutiful, modest nods to Tudor and Arts and Crafts, that I liked and admired. In my head, they are still inextricably linked to the welfare state, to free education and school milk, to Clean Air Acts and more equality than Britain had ever known. None of my father's family (he had six brothers and sisters) stayed on at school beyond the age of fourteen. They grew up in a tiny house in Hackney Wick, which the authorities later demolished as part of the slum clearance programme. Their children grew up in semis, and went to university.

So I don't know why, if I ever had to catch the Tube into town in the rush hour, I was overcome by near-suicidal despair, why I felt I was hurtling along tunnels to blankness. Possibly it was because I was secretly convinced that I was being educated for what my parents would have thought of, already anachronistically, as a nice job, something vaguely secretarial, in the City. East End girls wore smart clothes and got the Tube into the serious, hardworking side of town, where they assisted in the moving of money (not attractive to me, with my numbers problem). I remember, one summer in the sixth form, taking the Tube into town every morning to a temp job as a telephonist-receptionist at a small firm of accountants near Smithfield. I couldn't believe how empty the work felt, how hollowed out *I* felt, and I lacked the information to appreciate that there were other jobs in offices that might feel different. (My teachers at school, who had been at university in the late 1960s and early 1970s, were invariably socialists, hippies, or both, and thought working in the City was monstrous, sad, a sell-out, or some combination of all three.) My mood probably wasn't helped by reading Solzhenitsyn's *Cancer Ward* (I fancied myself serious and thought that meant you had to read depressing books) as the train rattled through Leytonstone and Bethnal Green and Liverpool Street, but I decided that if I had to do this for the rest of my life, if I even faced the prospect, I would just give up and die.

Charlie and I sat in front of Steve, this man I recognized but didn't, trying to do a good imitation of people who had been through the figures and knew what they were doing. The fact that we hadn't and didn't may have counted in our favour: we must have come across, I realize now, as fundamentally cautious about money, people who'd never previously taken any risks, who basically didn't have much nous. Charlie had grown up in a Methodist-influenced household where capitalism was regarded with a mix of awe and disdain. When he was planning to go to America during his gap year, his father not only organized a meeting for him with the bank manager, but accompanied him to it, as if to provide moral surety, proof of a background of fiscal restraint. Banks, in this world view, were places to which you went not so much for a service as for approval of your lifestyle; the bank manager was a kind of authority-wielding elder. That training in financial circumspection has stayed with Charlie, while my own caution was driven by straightforward fear. I was not financially adept, and had somehow achieved the remarkable feat of living in the Arabian Gulf for five years and returning with no savings. As a couple, Charlie and I had never played about with money, even in a small way, and our interest in cash had never got beyond a strong liking for shopping.

It helped, no doubt, that the economy was booming. The people at Atlas had suggested to Charlie that he might oversee any broadband ventures they backed in London, leading to an enhanced role and, one assumed, more of this money they seemed to have sloshing about. I was hazy about what broadband was, and Charlie may well have been too, being the sort of person who buys computers for their casings, but there was a good deal of talk about a 'lab' (a New Economy word for an office, presumably) to put the broadband ventures in, and he'd already done some preliminary scouting for property in Hoxton. In the event, there were to be no broadband ventures, but no one knew that then.

As we sat there, bandying figures with Steve, it became clearer and clearer that he thought it was a good idea, which was encouraging, because until then the only people who'd really thought it was workable were us and Elaine and Clive.

We had a fair amount of paperwork already, prepared in pursuit of earlier attempts at the money. Steve needed confirmation that Charlie really did have the contract with Atlas, so Charlie called them from the meeting and got them to fax over a letter of confirmation. I found this rather unseemly, slightly undignified – I was brought up to believe that discussion of money was vulgar (especially showing people that you needed it) and that you shouldn't embark on ventures that might be exposing – but then I have no money. This sort of modesty doesn't get you far.

Steve kept talking about 'bringing Charlie into the Barclays environment'. It was Charlie he was interested in, but that was fine by me. Steve said we'd have to get the land valued, which alarmed me because I knew we'd paid way too much for it, but he added blandly that we shouldn't worry: he'd get his mate Sean on to it. He'd worked with Sean for years. Sean was 'a practical man, like myself'. Would Sean be aware how much we'd offered? He would.

Basically, you could see, Steve was on our side. He not only dismissed my tentative suggestion that we could move out of our house in Hackney as soon as we got the mortgage on the land, but said he thought we might want to look at keeping the old house even after the new one was finished so as to rent it out. He had bigger ideas about this than we did.

Steve thought he could get us a bridging loan of £325,000, which would cost us a lot less than the one we'd been offered before, i.e. would be one that we could actually afford. So with our savings from the Atlas adventure, we'd be left having to increase the mortgage on Malvern Road by £225,000.

This didn't look like a problem: when we had put the house on the market months before, we'd sold it inside a week. Almost everyone who'd come to look at it had made an offer; we'd had three competitive bids and the winner would have paid cash, if we hadn't failed to get the house we were after and pulled out. And that had been months ago: it had probably gone up another 20 grand since then.

After the meeting, Charlie and I hugged each other outside on the pavement, slightly dazed and disbelieving. If it had been left up to me, I would have given up days earlier, weeks earlier, not only because the whole thing looked irrational, impossible, but because I had a hard time getting my head around the idea that I was the sort of person who might build her own house.

This probably sounds histrionic, not to say unlikely in this age when every other fifteen-year-old expects to win *Pop Idol*, but I was not fifteen. I was the mother of a fifteen-year-old, and of a different generation, and was definitely brought up not to get above myself. Even when I was in my late thirties and planning to marry Charlie, my mother was at it, anxious to see that I wasn't getting too cocky. I was planning, I remember, to buy another dress by Catherine Walker (the first one had worked so well) and grumbling about the price, when my mother commented, 'Well, they *are* dresses for princesses,' i.e. not, then, for the likes of you. She meant well, I know; she meant to ease my disappointment at not being able to afford the dress; I'd been over-ambitious in thinking I could have it in the first place. Princess Diana famously wore a lot of Catherine Walker and she and I clearly weren't the same category of thing. But the subtext was clear, to me, at least: I was not a princess. Not even when I was getting married.

My parents, I suppose, didn't want me to get bumptious, or be too disappointed with life – plus, I was a girl and it was incumbent on me to behave with modesty and discretion. Anyway, the upshot of a million subtle messages about not being a princess is that I have sometimes struggled to find a sense of entitlement, which has been all the more troubling for being allied with the arrogance of ambition. So, at the age of nine, I read an article in a children's annual entitled 'So, you want to be a journalist?' and I thought, 'Yes! yes, I do!' But I told no one, convinced they'd only reply: 'How about a nice job in the City? So convenient for the Tube.'

Anyway, luckily for me, I live with someone who has entirely healthy levels of self-belief, which he generously spills and splashes, like a fashionably brimming swimming pool, on to the people around him – children, colleagues, me – so we hadn't given up.

The worst thing to have emerged from our solicitor's search was that nobody knew who owned the lane. The council had never adopted it, which explained why it was so pitted and full of puddles and why there wasn't a sign at the end to tell you it was there. But roads of unknown provenance are a fairly common occurrence, and there's a standard procedure. You take out what is known as a defective title indemnity, which is an insurance, roughly to the value of the house, against someone's turning up at some point in the future and refusing to let you cross their land to get to your front door.

The only thing that was holding us up now seemed to be Lloyds' dilatoriness in sending a surveyor to our current house in Malvern Road. With three days to go we were hassling them to get someone over to us at the rate of roughly once every four hours. After everything, the idea of losing out for being late didn't bear thinking about. The people who were buying the house behind our plot had close friends who'd bid for the land;

I imagined them waiting eagerly for the deadline to expire so they could slip back in with their ready cash. As usual, I was over-reacting to authority figures (what does this say about me, that I am capable of seeing estate agents as authority figures?). In reality, I dare say the time limit was imposed mainly to find out whether we were serious before we wasted too much of everyone's time.

In the event, we weren't ready, but neither were the vendors, so the deadline passed without a whisper. They still had to produce something called a statutory declaration of occupation, an abstruse but vital piece of documentation, which I gathered would prove they owned the land and weren't just squatting on it. They had to get this, I think, from the previous owner, who appeared to be dead.

In the meantime, they came up with another stipulation. A wall was needed across the garden to separate off our plot. For some reason, we appeared to be the people everyone had decided should organize and pay for this. The couple who were buying the house and the bulk of the garden insisted on having a promise to build this wall included in their contract of sale. The vendors didn't want to pay for it, so they were making it a condition of sale to us.

Getting a wall built was fair enough, but I didn't see why we should have to pay for it. Certainly not all by ourselves. By rights, I pointed out to our solicitor, the vendors should pay, since they were the ones dividing up the land into parcels. So he put this to them, and they refused. OK, I said to the solicitor wearily, in that case we should split the cost with the other buyers, because the wall was for both of us. It was their suggestion, and, until our builders moved on site, it really didn't matter to us whether they had our nettles or not. So he went back and got the same response. Nothing doing: they wanted a wall of London stock brick, they wanted us to pay for it, and they wanted it inside a month. And they wanted all this to be written into the contract of sale.

Jan Morris once wrote that she never haggles when she travels, because the amounts involved are invariably less significant to her than to the locals, and because it makes her tense, wastes time she could be spending doing other things, and leaves her with a sour taste in her mouth. There are probably two types of person when it comes to haggling: those who positively relish it as a kind of exercise of emotional and intellectual muscle, and those who basically can't be arsed. I am in the can't be arsed camp, which is not a good thing: I have endured sleepless nights in hotel rooms as lifts cranked and creaked up and down the elevator shaft adjoining my bed. My sister – and this may explain why she is a successful businesswoman and I'm not – almost never accepts the first hotel room she is given, more or less on principle.

But even if I hadn't been essentially lazy, I was outgunned. The house behind was big and swanky; there was room to build on either side; the garden was wide and long (though obviously not as long as it had been), with mature plants, magnolia and eucalyptus trees and a lawn on which small children might drive plastic tractors and learn to play cricket. Even so, it was not a house to fall in love with. It was 1950s pastiche-classical and looked as though it belonged on a golf course in Surrey with Bruce Forsyth living next door.

There were always other houses. There were other houses, *exactly like this*, not so far away, in Chigwell, lived in by blonde women with enough money to re-do their roots every other week (women whom, when I was growing up, I was never entirely sure whether I was supposed to emulate or despise). But there weren't any other bits of land. Not with planning permission. Not in the place in which, if we'd had to choose, we would have chosen to live. From the vendors' point of view, the house was both more expensive and less saleable. People were queueing to buy our land and I wasn't about to lose it because of a wall.

I made one last attempt to be tough. I called our soon-to-be-neighbour, and said magnanimously that yes, we would pay to build the wall, but as we were making this gesture, sort of out of the goodness of our hearts, perhaps they could make a reciprocal one, and wait until our builders moved on site, which in my estimation would be in about five months, at the end of the year? No chance. He saw my point entirely, of course, that it would be much more expensive to get in separate wall-builders, and it really didn't matter much to him, but his wife wanted it. And she wanted it within the month. And that was that. I was dealing with a princess.

So there we were: outflanked and outranked by emotional royalty and faced with the first bill we hadn't expected: £7,000 for a garden wall, to be found out of current income, which was already under pressure, plugging the gaps left in our loans to buy the land. And we *still* hadn't got the extension on our mortgage from Lloyds, because they hadn't sent round a surveyor. We were three days past our deadline and he was an hour late for his appointment when he finally showed up: a scrofulous figure in stained Hush Puppies, with a grey canvas shoulder bag that looked as though it had been useful around the time of El Alamein, possibly actually *at* El Alamein, and a shuffling demeanour.

He had, he explained, got lost. This was not really good news. It meant he didn't know the area, which, for a surveyor, is like being a runner with no toes.

London is both socially stratified and not. Compared to most American cities, it's incredibly mixed, classy squares jostling cheap shopping streets (and nowadays, feeding off them for funkiness, allowing all teenagers to pretend they come from de ghetto). The locals, though, know precisely the point at which each of the classy squares fades into scruffiness; they can calibrate the stage of gentrification of each street, each terrace, with the exactitude of nano-scientists. This insider knowledge is what determines property prices – more, probably, than any other factor, and you probably have to be a native not just of the city but of the district to finesse the trends.

London Fields, where we lived, is an estate of a dozen or so streets of elegant houses thrown up speculatively in the 1860s to accommodate the growing middle class employed in clerical jobs in the City. The area had been substantially gentrified in the last twenty years, and has a good deal of charm: a number of the roads still had what were known locally as the backlands: overgrown alleyways between the gardens that were originally meant to house the mews, except that the railway had come before they could be built and left them to squirrels, foxes, brambles and trees. But London Fields is bounded on one side by Mare Street, which my parents remember as a well-to-do high street, with good Jewish dress shops where you were served tea as you tried on exquisitely tailored coats, but which has been a dump more or less ever since. Even in the period that has elapsed since mid-2000, when the surveyor came round, Mare Street has had an injection of regeneration monies and acquired a new library, a redeveloped music hall, other arts venues, good Vietnamese restaurants. But at that stage, the library was still operating out of a Portakabin, there were far too many slot-machine shops and, according to my children, people would issue out of the down-at-heel hairdressers in the evenings to offer stray passers-by (i.e. them) drugs.

On the other side of London Fields was Kingsland Road, already becoming very fashionable at the bottom end, where it filtered into Hoxton and Shoreditch, and not that bad on Kingsland Waste, closest to us, where there was a Turkish greengrocer who sold honey off an Anatolian mountainside, and Ashok Patel, probably London's handsomest newsagent. But only a couple of hundred yards up the road lay Dalston Junction, an unpleasant crossroads featuring stalls of shoddy clothes, a hideous mini-mall purveying tat to the sad-eyed asylum seekers from the nearby sweatshops, and shops selling everything-for-a-pound. The traffic seized up here, and you had to walk quickly to avoid being fatally choked by the fumes.

So if the surveyor had got lost, the chances are it wasn't in a particularly nice place. He looked nervous, and probably thought he was about to be shot. He wandered around the house and I trailed after him, keeping my distance on account of the bits of him that were flaking off.

Had we had the house valued before, he wanted to know, so I launched into my explanation about the near-sale and the sealed bids and getting more than the asking price after only a week on the market.

'What's the area like?' he asked, in a voice that suggested he was really saying: 'But that can't be right: the area's crap.'

'You should speak to local estate agents,' I told him. 'Currells in Islington are good.'

'In Islington?'

'They handle most of the sales round here.'

'What about local estate agents?'

'They are local,' I said, slightly desperately. 'Or Holden Matthews. They valued it too. Higher, in fact, than Currells.' I smiled, apologetically. 'They're in Islington, too.'

He went into another room, then drifted back. 'How long have you been here?'

'Eight years.' I paused, considering how eager to appear, then thought, sod it. 'It's changed quite a bit in that time. The knock-on effect of Islington, and then Hoxton and Shoreditch.'

He looked at me blankly, as if these were places he had only dimly heard of and of whose geography he was uncertain, the Burkina Faso and Surinam of North London.

Charlie reappeared and asked how long he thought it would take to get his report to Lloyds.

'I need to look at other property prices in Hackney. Average prices. And then I'll write to them.'

'Right,' I said, not wanting to appear too anxious, 'but obviously, Hackney is rather *varied*.' I meant that he'd better realize this was a nice part.

'I don't know if they explained,' Charlie said, 'but there is some urgency about this . . .' We'd just heard we could exchange the following day.

'Oh, yes,' the surveyor smiled bleakly, in the manner of one both put-upon and disapproving, and shuffled off in a little cloud of dust to the front door.

Charlie looked at me worriedly. 'I didn't like his expression when you told him how much we'd been offered for the house.'

'Maybe he didn't believe me?'

'Almost as if he thought a house couldn't sell for that much.'

'Or not this one, anyway.'

Charlie called Lloyds several times the following morning, because we had an appointment with our solicitor at 2 o'clock to sign the papers for exchange. After all that we'd been through, exchange was now all it was to be. Despite the strictures at the outset, the other side now wanted a couple of days before completion. Still, even we, cavalier as we were, couldn't exchange without having secured the money.

The people at the bank promised they were hassling the surveyor and his report would be with them soon. But we should stop worrying: they didn't anticipate any problem, since from all we'd said the house was worth plenty enough to get the money. They'd get back to us as soon as they had the go-ahead.

We were at the solicitors' office on the edge of the City, waiting to go into the meeting room, when the call came through on Charlie's mobile. We'd got the loan. We could have an extension of £205,000.

It was £20,000 short of what we needed.

We'd expected £225,000. We'd hoped for a valuation of £430,000, which was £20,000 less than we'd been offered and had accepted months ago. It meant the surveyor must have valued the house at £400,000.

Charlie whispered to me to go in with the solicitor ahead of him. He looked grim, and he rarely does this, so it was quite scary. He is one of the most equable people I have ever met: I have never, for example, seen him heated in political discussions; he lays out positions calmly, with detachment, and doesn't approve of getting aerated about abstractions. But like a lot of calm people, when he flips, it's frightening, because he flips with force. I sat smiling vaguely at the solicitor and making airy small-talk, as if nothing untoward was happening, while loud shouting could clearly be heard from the ante-room.

Charlie had called Directory Inquiries, got the surveyor's number and called him in his office in Southgate. Had he, Charlie demanded, spoken to local estate agents? The surveyor admitted that no, he hadn't, but he'd looked up property prices in Hackney. Charlie yelled at him that he was incompetent, a disgrace to his profession, that he hadn't paid any attention at all to the house but had proceeded according to strange and distorted assumptions of his own, and rudely hung up. To compose himself and try to feel better, he envisaged the surveyor in a cobwebby office, surrounded by grey filing cabinets with stiff drawers and old files that no one would ever examine again, the Miss Havisham of chartered surveying.

Then he returned to the room. Unfortunately, even after the shouting episode, we were still short of £20,000. With a sudden thrill, I realized I had £13,000 in my current account, saved up for tax and VAT. Charlie left the room again, spoke to Lloyds and established that he could get the remaining £7,000 on overdraft. I felt enormously smug: after all the to-ing and fro-ing for money, none of which had had very much to do with me, I had made it all happen. I was crucial.

(Of course, this is mad, because Charlie and I are married. But this is the warped way you can easily start thinking if you earn less than your partner.)

At the last minute, another solicitor came into the room to witness the signatures, which was the first indication I'd had that our solicitor wasn't in fact a solicitor, but a legal executive. But it didn't seem to matter. He'd done our conveyancing on Malvern Road before this, he'd been quick, and he'd solved the problem of the unadopted lane with the indemnity insurance. We signed and, two days later, we completed.

On 30 June 2000, we became the owners of a patch of weeds and an enormous amount of debt, including a bridging loan which would cost us £87 for every day we had it.

3

We knew very little about Joyce and Ferhan beyond the fact that one was American and the other was Turkish and Hugo had once said something to the effect that what he'd *really* like to do next was build a house from scratch, and he knew he could do it because he had a great architect. This remark had lodged in my head: possible to build from scratch, have good architect.

Plus, we liked sitting in Hugo's kitchen, and there had been the clever plan for the subterranean offices at Malvern Road . . . but that was pretty much it. We didn't know how a woman from Chicago and another from Bursa, a town in Turkey we hadn't heard of, had ended up in Islington, let alone whether they were capable of building a house. They'd never done one before.

In *Grand Designs*, the book-of-the-series, Kevin advises, 'In the first instance, follow Jane and Gavin's example, and interview several architects.' You should, he explains, ask them about their history, their other projects, their working methods, which architects have influenced them, their *thoughts* about buildings.

Had we done this with Joyce and Ferhan, we might have discovered that Joyce admired a dead architect called Carlo Scarpa, but, given our own ignorance, it wouldn't have helped much. Ferhan might have asked, as she often does: 'Do you think we're lesbians? A lot of people think we're lesbians.'

Well, no, she probably wouldn't have, not at a first meeting. But what might we have discovered to persuade us that they were right for us? They were both young mums (so not, apparently, lesbians). Their children went to the same school as Harry, our five-year-old, which I didn't altogether like: I wanted them to be more radical, more *out there* than us. (I don't know where I imagined their children should be educated – some sort of super-creative, innovation-blazing state school that everyone would be clamouring to get into if only they were hip enough to have heard of it, I suppose.) And neither of them had set out to be an architect. Joyce had begun by studying veterinary medicine and flirted with accountancy before taking a drawing course and deciding that sitting with the windows open and listening to music was preferable to fretting over figures or inserting suppositories into farm animals. By her own account, she didn't particularly distinguish herself at university: 'I had one professor who said to me, "What are you doing here?" and I said, "I don't know." And he said, "You will see: architecture will be your life," and I just looked at him and thought, "Architecture's never going to be my life, honey, I got better things to do."'

Ferhan didn't spend her childhood looking at buildings and thinking she could single-handedly take architecture into the twenty-first century either. She thought she probably wanted to be an economist or a journalist. Her father was an intellectual – 'looked like Jean-Luc Godard, smoke curling up in front of him' – who read Chekhov for her bedtime stories and ran the Cinémathèque in Bursa, so that she was steeped in European cinema and saw Vittorio da Sica's *Bicycle Thieves* when she was seven. As it was assumed she would, she won a scholarship to a good school, where all lessons were taught in English: 'I was brought up by this bunch of intelligentsia and there was nothing to do but go to university and be yourself,' she says now, 'so I rebelled when I was sixteen by deciding that I was going to marry the local shopkeeper. He was fifty and a cripple.' She was with him for eighteen months, and her grades imploded.

Eventually, once her father started letting her bring the elderly neighbour home, she dumped him and caught up with most of her work, but not enough to secure the university place she should have had. And then she seemed to have had some sort of existential upset with the application form. 'I ticked economics. And chemistry, because by then I was seeing a younger friend of my father's who was studying chemistry at Ankara. And I put architecture in there, but it got mixed up because of some rubbing out and I didn't even apply to the best architecture school.'

Accepted on to the architecture course (the inferior one), she hated it and almost dropped out. After a lot of persuasion, her father convinced her to stay on in Istanbul and apply herself to architecture, instead of merely to boyfriends. She engineered a switch to the school she should have applied for in the first place, where she started coming top of her class.

Joyce and Ferhan's story is not one of single-minded ambition and willed destiny. It isn't, in other words, a man's story. They weren't in Britain because Britain is a particularly good place for architects to be; they ended up here for the reason that women have traditionally ended up anywhere: men. Joyce had spent a year in Rome as part of her degree, where she not only discovered Scarpa, but also another, English, architect, Bill, whom she subsequently married.

Ferhan was offered a job in her professor's practice after university. Three years later, she was asked to do some speculative drawings for a project his son was working on with an English property developer called Erik Pagano. Ferhan went for a drink with Erik, fell in love, and married him.

By 1988, Ferhan was in London and working for a big commercial architecture practice that specialized in multi-million-pound projects and had a staff of 120. One of the 120 was an American woman, open and easy-going and always laughing, and Ferhan says she thought, 'I'll be friends with this woman: she's fun.' She and Joyce knew almost nothing about each other's architecture because they never worked on anything together and, in any case, the work they were doing was all detail, joining the dots of someone else's designs, and left no room for individuality. Joyce hated it; she went back to Florida. Her entire family (she has four brothers) had decamped to the state from Chicago, and life was easier for her there than in Britain. She found interesting work with a very good practice. But Ferhan wouldn't stop pestering her. 'She kept calling, saying "Come back, come back," and endless letters – I mean, these days who writes letters?'

Luckily for Ferhan, Bill didn't like Florida. He couldn't settle. Quite soon, he realized he had no desire at all to live in America. Joyce's prospects, on the other hand, were better than ever; she had offers of work coming in from all over the country. But when Ferhan was headhunted by an expanding practice in London and wrote yet another letter demanding that Joyce come and join her, Joyce caved in to the pressure and decided that she might as well. Almost immediately, she and Ferhan started working on side projects of their own in their spare time.

'What, you went into business without knowing anything about each other's architecture?' I asked them incredulously, much later.

'We sort of did,' said Ferhan.

'We did, sort of,' said Joyce.

This meant getting up at 7 a.m. and cycling into the little office they'd rented from a woman Ferhan had met at the gym. 'Then we would run, literally run, to be at our desks by nine. Then at lunchtime, we would leave separately, five minutes apart, and run round to our office again.'

'We were stupid,' Ferhan says now. 'Why would anyone care what we did?'

The office was tiny and airless; what with their bicycles and the gas fire, 'We couldn't move without hitting each other on the shins.' One day Joyce smelled smoke. Ferhan said that was OK, she was smoking, which she was, but mainly because she'd caught fire.

They won a competition, prompting the owners of the

restaurant Kensington Place to ask them to look at designing a brasserie on the other side of Kensington High Street. They answered an ad in *Loot* for a new office; when they said they were architects, the landlord said, 'Oh good, I need some architects,' and got them to work on designs for a nightclub. Then came projects that were actually built: a friend of Ferhan's virtually ordered her old boyfriend to use them on the refurbishment of his Georgian house. He made so much money doing it up and selling it that he bought a flat on several vast floors behind Upper Street and employed them again. Joyce, meanwhile, rented a house in Waterloo, where she met the owner of the bakery Konditor and Cook, for whom she and Ferhan have subsequently designed three sleek shops. Isabella Blow was another neighbour: they did her house; that led to a commission for Alexander McQueen's shop off Bond Street. By the time we met them, they had worked for photographers, hairdressers, designers and the editor of the *Spectator*, later the Conservative MP for Henley, Boris Johnson. (They were always winding us up about who would need more metres of bookshelves, us or the Johnsons. In the end we had 65 metres, to, I believe, the Johnsons' 68. But I like to think that we let them win, and that we are just more discriminating about getting rid of stuff that we're never going to read again.)

They had, in other words, some stylish, interesting and diverse clients. But they were still largely unknown. If you're going to go to all the bother of building a house, you'd imagine that you'd want to employ the best, most exciting architect you can afford: you're not, presumably, looking for something pedestrian. Joyce and Ferhan had some photographs of what they'd done, but architectural photographs only give a limited, glossed-up sense of what it's like to inhabit a space. They hadn't won any prizes for their work, or been much written up in the architectural press. And even if we'd asked them (which we didn't), I don't think we'd have got much sense of their *thoughts* about buildings because they don't articulate architecture, they just do it. They actively dislike what they call intellectualizing architects.

They do, obviously, have a style, which is rooted in modernism, is pared down without being rigidly minimalist, relies on the use of a limited palette of materials on any particular project, and employs certain tricks, or devices – surfaces that move from inside to out, benches. (I don't think they've ever done a project that hasn't ended up with at least two benches. We have four.) But I've only worked this out for myself over time. I don't think they would ever say it themselves, for fear it might limit them. They talk – both of them – as if talk might be rationed tomorrow, but words aren't really their medium, they're not how they *think*.

So they were an unknown quantity. Yet we were expecting them to understand us, to translate us into materials, to body us forth in all our complex emotions and diffidences and enthusiasms. We trusted them to build a house that we would love. Looking back, I think we were mad.

The weekend after we secured the land, we decided we should have a celebration, so we invited Joyce and Ferhan to the land for champagne, plus Hugo and Sue, who had introduced them to us, and Elaine and Clive, who had offered to give us their house, and everybody's children. Our family drove to the land en route from Freddie's school open day, and the car was full of stuff he'd been keeping there and was bringing home for the summer. Sometime before we arrived, I squeezed myself in amongst this junk to feed Ned a pot of something or other for his tea. (I'd like to say this was homemade, softly sweated carrots perhaps, tenderly pressed through a sieve, but he was my fourth child so it would have come out of a jar. Henrietta, on the other hand, had no jars in her infancy but so many sieved organic carrots that she contracted something called carotene anaemia, which turned her yellow, like an ageing roué with hepatitis.)

We parked in the private road, risking the clampers. I was

trying to pass Freddie's trainers to him in the front, unhook Ned from his car seat, avoid getting the food that was down his front down mine and climb down from the people carrier, when I somehow dislodged the lid from the plastic tank at my feet. The plastic tank contained a praying mantis, although not for long, since it saw its opportunity and lunged for freedom. Seeing that it was about to escape, I shoved the lid back on, hard.

There was a distinct crunching sound. I looked down. Half the mantis was inside the tank. The other part was hanging at an awkward angle over the edge. Instinctively, even while gasping at the awfulness of what had happened, I lifted the lid a fraction and surreptitiously tried to shove the dangling portion of mantis back inside.

'What are you trying to do?' asked Freddie, aghast at the sight of me fumbling with his decimated pet.

I looked up at him numbly.

'Were you trying to put it back in the tank?' he wailed. 'When you'd killed it? How could you?' He gazed at his squashed mantis. 'When I've been looking after that for *months*? When it's doubled in size!'

I was mortified. How could I? What kind of mother was I, to kill my son's pet? I wanted to say that I'd replace it, but knew instinctively that this would be a bad offer at that moment. Freddie, still in his stage makeup from *Excerpts from the Pirates of Penzance*, and already 5 foot 8, stood weeping inconsolably in the street. There was champagne on ice in the boot and, over the wall, half a dozen people waiting to celebrate.

'I'm sorry, Freddie,' I said again, helplessly. 'It was an accident.'

'Oh,' Freddie sniffed, 'so you didn't do it deliberately?'

'Mum, how *could* you?' said Henrietta, turning up at that point with her boyfriend, Matthew.

'It's not even dead!' Freddie wailed, inspecting it more closely. 'You've cut its legs off! It's only in agony!'

Freddie carried the mantis mournfully on to the land and

crushed it (although I have since read that mantids can regrow legs, so we may have been too hasty. They are surprisingly robust for animals with so many thin bits. Insects generally are alarmingly tough. Later that year, one of Freddie's stick insects escaped from another tank in his bedroom and was presumed killed by the cat until, months later, it leapt from the curtains on to a child's pillow during a sleepover).

'It's like the whatsit farmers in Ghana, or somewhere,' I said half an hour later, trailing across the grass to where Freddie was still communing with his squashed insect. (I was slightly drunk.) 'Nabdam. When they want to build, they invite their friends and family round to the site and kill a chicken and throw it on the site, and if it dies with its beak in the air, it's a sign the ancestors approve.'

'Yeah? Well, my mantis was facing downwards. One half inside the tank, one half outside.'

So if we'd been Nabdam farmers and the mantis had been a chicken, we would have had to find another site. It's important that the ancestors approve.

And when I was less upset, and less drunk, that didn't seem to me such an unreliable way of looking at things. What, for example, about the people who'd lived in the building on our site that had been bombed (did it take a direct hit? Were they in it at the time?): mightn't they have wanted to know how it was changing, what plans we had for it? Memories are mixed up with, embedded in places: the streets of Hackney Wick in which my parents and my aunts and uncles grew up were knocked down in the postwar slum clearances to make way for tower blocks and motorway extensions, flyovers and dual carriageways. But they still – those of them who are alive – like to go back and trace the patterns of the streets under the concrete, to recollect their childhoods. The places reassure us that the experiences were real and not imagined, just as novels in which a sense of place is powerful rarely feel untruthful.

The time that I feel most furious that my father is dead is when I pass the little alleyway in Clerkenwell where his office used to be (which, in fact, I do very often, as it's on the way to Harry's school). In his day, the area was full of businesses like his, where bits of electrical equipment, or perhaps clocks and watches, were taken apart and put back together by men in brown coats. The windows were grimy, often barred; even the shops – grim stationers selling manila envelopes out of dusty boxes – seemed to turn their backs to the street. You had to walk half a mile to get a sandwich made of limp bread and margarine, filled with something that had been sitting too long in a metal tub and was congealed on the surface.

Now there's a Pizza Express on one corner of his alley and a Prêt à Manger on the other, with a Starbucks next door. I wonder what he would have made of this: if it would have amused him that Clerkenwell, of all places, has become a young persons' place, its factories littered with expensive lofts, or if it would have made him say something sceptical – he was generally inclined to scepticism – about the predictability, the shiny uniformity, of the changes brought about by globalization. I can scarcely pass the alleyway without giving the thought a glancing nod, and, in that sense, he haunts the place.

The land was looking smaller each time we visited. Possibly literally: someone had hammered wooden posts into the ground a couple of feet in from where we thought our boundary should be (i.e. where the estate agents had said it would be, and where the garden got scrubby, which also happened to line up with the boundary of the properties next door. There were, to make the whole thing clearer, a couple of concrete posts on this line, which we assumed must be the remains of some previous wall).

'Well, I hope they're not meant to represent some kind of boundary,' I said, cheerfully, loping around them with my champagne glass.

'Nah,' said my sister. 'How could they be? They don't line up with next door. Anyway, your land is the unkempt bit.' She lowered her voice. 'Are your architects lesbians?'

'Elaine, they're business partners. And they've got three children here between them.'

'That doesn't mean anything.'

'You just think they're lesbians because it's unusual for women to be architects and set up a practice together. That's incredibly sexist. Do people think you're a lesbian because you run a successful production company?'

'It doesn't matter, of course,' said my sister infuriatingly. 'But I bet you they are.'

Before we left for a week's holiday, I asked our solicitor to investigate the sudden appearance of the posts (which, when I was sober, were somehow more worrying). We set up a meeting with Joyce and Ferhan for when we returned; in the meantime, they would arrange a site and soil survey. If we'd been sensible, we would have done this before we paid much too much for the land: what if it contained an old toxic waste dump? A new toxic waste dump? High concentrations of radon gas? Killer cables?

Across the lane was another site, already sold when we first saw our plot – again, for what seemed like a ridiculous amount of money, since it was smaller than ours and an even less promising shape. Work had already started when we first saw it – pilings had gone in – and then abruptly stopped. We were told that a workman wielding an electric drill had gone straight through a major electricity cable and been thrown to the other side of the site. He was lucky not to have been killed. When I eventually met the owner, an architect who was building for himself (over, it would turn out, many years), he told me that if he'd known that he was putting his house on top of an electricity substation, he wouldn't have bought the land. His house is still not finished.

But he hadn't had time to investigate properly, and neither had we; our site and soil surveys together cost £2,831.70 and

I'm not sure we would have wanted to spend that on a hope. Now, though, we owned the hope, and had to.

The last thing I did before we went away was send Joyce and Ferhan a brief. This called, effectively, for a tardis:

5–6 bedrooms
kitchen/breakfast/family room
2 studies – one that we could have a bed in?
2 bathrooms
utility room
shower room

As well as this, I put down:

Masses of light
Lots of glass
Limestone floors

Then I elaborated some random thoughts:

The kitchen to be the largest room and heart of the house: the room we will really live in. Facing on to/integrated with the garden. We will always eat here – the dining part ideally next to the garden. The kids will probably spend most of their time playing here.

Reception room – another place where people can go to watch television/play music/talk to friends. Not forbidding but not a den either.

Henrietta's bedroom – to fit a double bed

Freddie's bedroom

Harry and Ned's bedroom/s – should they be together or apart? Will there be any space for the boys to play other than the kitchen/family area?

Need lots of storage space – for toys, camping equipment, bicycles, scooters, roller blades, etc., clothes

Should we consider an L-shape – with studies/adult bedrooms on one wing? [Charlie put that in; I thought it was getting above ourselves to suggest what shape it should be. But he had a point about getting away from the children, so I left it.]

Geraldine wants to garden, so light into the garden is a consideration: we don't want to block out the sunshine.

Geraldine's study – doesn't have to be large.

I don't know why I put that last bit in, unless as an acknowledgement that we were asking rather a lot of a small (and shrinking) site, and perhaps because I am my mother's daughter. My mother, when asked over for dinner, will say: 'Don't worry about me: I'll be happy with a few scraps.'

Much later, Joyce and Ferhan were talking about what makes a good client and said it was someone who gave a very clear brief and then left them alone. 'Some clients go away for the whole design period,' Ferhan said wistfully.

'Our brief was a bit sketchy, wasn't it?'

'Nooooo,' they replied unconvincingly. 'Anyway,' added Joyce, 'we'd already done those drawings for your other house, so we knew quite a bit about you.'

But they didn't, not really. They didn't know why it was important to our family that we'd lived in the same house for eight years, how rooted we felt in it, and why. They didn't know about Henrietta and Freddie's father having married an MP, or that they divided their time between a series of houses in London and her constituency in Redcar. The children had loved Redcar when they were young: the house was big, solid, capacious, tatty round the edges, overlooked the iron sea. But it wasn't where they went to school, or where their friends lived; people spoke differently and treated them as special because of their connection with the MP. There was black coal on the beach, washed down the estuary, the shops on the esplanade were decaying; it was a place that seemed to be struggling to stay attached to England, to be crumbling off its North-Eastern shore, held on only by the bulk of the chemicals factory at the end of the beach.

They moved around in London, too, depending on the vagaries of their stepmother's being in power and out, a backbencher and a minister: a flat in gloomy Dolphin Square, lugubrious and peopled by deracinated MPs; a soulless grace-and-favour house in Eaton Square that they all hated, in spite of the grand address; another government place in the attic of Admiralty House, overlooking Horseguards Parade and over the head of John Prescott. And, in between, spells in Southwark and Islington. They whirled around in this vortex of homes, liking the novelty and the drama – besides, how many teenagers get to live just off Trafalgar Square? But there was also something feverish, convulsive about it. And compulsive: being peripatetic came close to an addiction for Henrietta: when she was fifteen, she said she felt uncomfortable spending more than two nights in the same bed.

When their father and I split up, my cousin, who is a divorce lawyer, cautioned me gently that the current thinking is that children need one home. But it was an impossible choice: he and I both wanted to be parents and we both believed that the only way to do so is to be involved, and involvement seemed to be measured in time. So we split their time. Henrietta and Freddie had three homes for most of their childhood, and Freddie's PE kit was always in the wrong one.

We didn't think we were dividing them in two, like the fake mother in the Solomon story; but now I suspect that perhaps we were. They went to the schools they would have gone to anyway, but some evenings they were collected at 6 o'clock and driven to the other side of London to sleep. One weekend in two they couldn't do music and pottery classes or see their friends, because they were on their way to the North. They moved around more than a lot of traveller-children, just in nicer clothes.

I didn't blame their father for this, really, or not beyond that lazy way that you like to blame people who aren't there to answer back. I acceded to it, not knowing what else to do, what the preferable alternative might be: to suggest that he shouldn't see them as much? That I shouldn't? And it was so painful that I was convinced it must be doing *someone* good. I remember once dropping them off at King's Cross and seeing them go off with their other parents, looking like a happy family, and wanting to lie down on the concourse and wail. It took a monumental effort to move one foot in front of the other and get out of the station. And even when I fretted endlessly about the fact that they slept in the back of the estate car on the motorway on Sunday nights – did they have their heads towards the back; what if something went into the car? – I told myself that this anxiety was less to do with their vulnerability than mine. I rationalized my worries – and maybe even genuinely believed this is what they were – as a reflection of my unhappiness about the whole thing, a way of punishing myself for having let it happen at all.

Through all of this, Charlie and I lived in the same house in Hackney – which, initially, might have been just another one in the roster of homes, although time had made it something else: a fixed point, at the very least; a place you could rely on if only because we'd gritted our teeth and stayed there for ten years, determined to believe it was capable of accommodating everybody.

As the elder child, Hen had borne the brunt of the shuttling childhood. When the official driver turned up to take them across London, she was not supposed to lapse into helplessness. Since Freddie had quite clearly abdicated all responsibility for his PE kit, people expected her to keep an eye on it. Not altogether surprisingly, as soon as she had much choice in the matter, she started opting out. Giddied by the St Vitus's dance of homes, she began to withdraw from the rackety arrangements, to find she'd 'got stuck' at her friend Tess's or later, at Matthew's. At times she actually seemed to prefer their families to her own, and, perhaps, to prefer being a guest, an observer, to a wholehearted participant. In our house, she claimed, she couldn't hear herself think for all the children – which I suspect meant she felt the house wasn't really about her.

She had taken the supposition that she might have more than one home to its logical conclusion. Now she could only assemble her sense of homeness, derive her sense of security, from a network of houses and families.

Then she started sixth form and something (possibly not unrelated to the end of my nauseous, bad mood-inducing pregnancy) changed. Hen took a decision that, to work, she would need a base, and unilaterally overruled her parents' questionable, if benignly meant, arrangements. She saw her father, to whom she has always been close, as often as ever, but she located herself primarily with us.

For months and months I had wanted Hen to come home, and now she had done so. In spite of her father's and my best efforts to deprive the children of home, home was what our house had become.

Leaving Hackney would be a huge dislocation. It had to be worth it, not just for Charlie and me, but for all of us. Joyce and Ferhan could have no conception of how onerous a responsibility they'd taken on, of how much baggage I was dragging with us; nor that when I said, 'Kitchen to be the largest room of the house,' I meant all *that.*

We returned from our week away to a bundle of correspondence: letters from our solicitor to the estate agents and the vendors' solicitors, plus replies. The only one that really mattered was the faxed note from the vendors, which pointed out that the deeds clearly stated that the plot was 75 feet by 55 feet (which they did) and revealed that the wooden posts marked the place where the 55 feet ended. The letter concluded: 'We are sorry if the purchasers imagined the concrete posts defined the plot, but they are not 55 feet from the southern boundary wall' – i.e. the purchasers were prats for not measuring.

Why, I stomped around asking myself as I lobbed dirty clothes at the linen basket, would our land start a yard into the scrub, and not line up with the boundary next door? Who had decided anyway that the boundary was 55 feet from the southern wall, and when? We now knew the two plots had been bundled together before the current owners had bought them, and perhaps for a long time before that. Had the deeds always stipulated that, if separated, the southern end of the site should be 55 feet deep? Or had somebody just *made that up*?

I wrote to the Land Registry, inquiring about the history of the site(s), but, although staffed by very friendly people who give every impression of wanting to help if only they could, the Land Registry is more secretive than Porton Down, the Atomic Weapons Research Establishment and GCHQ rolled into one. I don't know why this should be, unless it's something to do with a particularly English attitude to class, whereby owning land makes you more important and gives you more rights than anyone else, including the right not to reveal what land you own. The nice members of staff at the Land Registry were not at liberty to open the file. Anyway, as people who cared about me started gently pointing out, it was irrelevant now, because 55 feet was what we'd bought.

'The good news,' the solicitor said brightly, 'is that, technically speaking, the wall should start at fifty-five feet and come on to your land, so you've got the whole of it on your side, but we've got everyone to agree that fifty-five feet should be the mid-point.' It was a measure of how neurotic I'd become over the whole thing that I was prepared to regard this as a triumph.

Still, we had something to look forward to: our concept

meeting with Joyce and Ferhan. One sunny Tuesday morning, we rang the bell of the scruffy door in an alleyway behind Upper Street and were buzzed inside, where we were faced with an improbably steep staircase. Their office was under the eaves, half a dozen desks pushed together at the front of the room by the windows and, at the back behind a long white cupboard, a meeting space around a wide white table. Everything, in fact, seemed to be white.

While an assistant prepared coffee, they gave us albums of photographs of their previous projects; I looked through them stupidly, not really knowing what I was looking for, or how to judge them, a little fearful of not liking them. I felt nervous, excitable, wanting to get on and see what sort of a house a person like me ought to be living in.

The first and biggest question, they explained once the coffee had arrived, was where to put the house on the site, and which direction it should face. From this everything else would follow, because it would dictate where the light came in, for how much of the day. Joyce, who was managing our project (they designed as a team, but one or other of them took responsibility for dealing with client, contractor and any other management issues), had already been in touch with Islington Planning Department to see whether there might be any constraints on positioning. There were; and the restriction was massive and intractable. The Department expected the new building to follow the footprint of the two developer's houses for which planning permission had already been granted.

'They said something about the Residents' Association,' Joyce said vaguely.

The developer's houses (as we referred to them, to show that they weren't in *our* league) were modern two-up, two-downs with a third bedroom in the roofspace. They had frontages on to the lane and small, north-facing gardens at the back. No doubt if they'd been built they would have been perfectly all right, but on the plans they looked mean. I thought of them as cringing, their purpose mainly mercenary. Many months later, I tracked down the architect who'd designed them. It was clear from our conversation that his client had always seen the real money as coming from the development at the other end of the site, where he had permission to knock down the 1950s villa and put up a block of flats. The houses in the lane weren't exactly an afterthought, but they were a bit of a bonus; they certainly weren't a labour of love. And even I, knowing nothing about how you should begin designing a house, recognized that given a virgin site you wouldn't put the garden to the north.

So Joyce – who'd been in dispute with this particular planner on a previous project – asked whether there were any circumstances in which his stipulation about the footprint might be relaxed. 'Only,' he said, raising his eyebrows, as if this were highly improbable, 'if the house were of outstanding architectural merit.'

This, quite clearly, was a mad thing to say. They were architects. What on earth did he think they were about?

The upshot, Joyce said, was that she and Ferhan were going to show us the concept for the house they wanted to build, and we'd have to make a decision about whether we wanted to take the risk to go ahead and design it, because it was conceivable that we could get all the way to planning and have it thrown out.

Positioning the house was difficult for other reasons, besides the Planning Department's intransigence. First of all, there was the huge ash tree dominating the north-western corner, with a preservation order on it. This was more than just a tree-in-the-way technical problem. The tree had been there for a long time: it was easily more than 100 years old and had flourished because it had been planted, or had seeded itself, in that precise place. In some sense, the site belonged to the tree. Weird and mystical feelings about trees aren't (it seems to me) just a product of contemporary worries about global warming and acid rain and deforestation and logging. To Hindus, neem trees have always been objects of worship. Christians bring fir trees into the house in winter (really a pagan thing, but anyway). In Shakespeare you only need a few trees for everyone to start changing sex and falling in love with the wrong person.

And the siting of houses in relation to trees has always been important. Any self-respecting English parkland, good enough to feature in a television adaptation of Jane Austen, for example, requires a few handsome and well-positioned oaks. Chinese villages often have a grove of trees or bamboo behind and a pond in front to ward off evil influences. The ash tree gave our site an identity, contours that we couldn't rub out.

There were other constraints, too: although the 1950s villa was some way off, the Planning Officer at Islington Council warned that anything that overlooked it might have difficulty getting through. And we didn't particularly want to look out in the other direction, over the scruffy lane. Lastly, as the Planning Committee had made clear when they'd approved the developer's houses, and as the Department reiterated now, any further development should respect the roofline of the mews (which had subsequently become a factory making glass and was now derelict) at the dead-end bottom of the lane.

Joyce and Ferhan turned over some pieces of paper to reveal a series of prettily coloured sketches. The house they envisaged was a long glass box, with more or less blank walls at either end. When you entered from the lane, the hall would rear up to the roof along its whole length. The right-hand side would be glass, overlooking greenery, and the roof would also be glass, on to the sky.

All the rooms would be to the left, looking out and opening on to the garden. There was a sketch of a little orange smiley-faced sun moving round the building, showing that the rooms would always be light and there would be plenty of sunshine in the garden, which would face west. From here, the building would resemble a doll's house – a long box with a glass frontage, capable of opening up along its entire length, with rooms on view. There was a sketch showing Harry in one of these and Henrietta and her friends in another, with a speech bubble asking, 'Why have these architects installed smoke alarms in my room?' while they presumably smoked a joint or cigarettes. At the edge of the kitchen/dining room/den which dominated the ground floor, there would be what Joyce described as an indoor/outdoor eating space, behind a metal grille. I didn't altogether grasp how it would work, but I loved the idea of eating outside even when the weather was a bit rough. By means of architecture, you could, apparently, eat outside in the winter without feeling the cold.

And then finally there was a prettily coloured impression of our bedroom, with a kind of courtyard in the middle of it with a retractable roof, so that Charlie could indulge his enthusiasm – the perversity of which everyone politely ignored – for outdoor washing.

It was dazzling. Or rather, bits of it were dazzling: the dramatic glass hallway, the indoor/outdoor eating space, the courtyard bathroom. It was difficult to take everything in, and these were the things I absorbed, because they were the most exciting. In the process I cheerfully overlooked the fact that Henrietta's bedroom was on the ground floor, next to the garden, and that the studies had been stuck on the roof in a funny kind of box, like a little hat. I was so taken up with the thrilling parts that I didn't bother to find out what Joyce and Ferhan thought the house would be made of. I did, however, in a masterly stroke of architectural priorities, ask what sort of plants they were thinking of for beyond the glass hall wall.

The question was, were we prepared to go ahead and work this up? It was a risk: if we got to planning only to have it rejected, we'd be back to square one. We would have wasted months. Ferhan explained that she and Joyce weren't saying that they *couldn't do* a house turned round in the other direction, to face the lane, but it would have to be quite different – perhaps involving light coming down into the middle, through a courtyard or something. If we were reduced to going for that option, it would have to be a whole different house (meaning that we'd have to pay for the design process all over again). They just thought this would be better. So they felt duty-bound to show it to us. And they made a proposal: if we wanted to work up this idea, they'd charge us an hourly rate to design it up to the planning stage. If it got through, we'd incorporate the fees into their overall percentage. If not, they'd charge us half the money.

It wasn't a difficult decision. On the basis of four drawings in coloured pencil and a sketchy floor plan, we were already in love with their glass doll's house. We didn't know what the alternative might be, only that it wouldn't be as good and, if we were supposed to be designing the ideal house, the beginning was hardly the moment to compromise. True, we had the phenomenally expensive bridging loan, clocking up the hundreds with every passing couple of days until we got to planning permission, but how long could it take? Three months, we thought, at most.

4

'Beginnings are delicate moments,' the architecture critic Witold Rybczynski has written, 'the beginning of a building no less than the beginning of a friendship or a marriage.'

He's right, I think, about friendship and marriages. If I hadn't turned away from Charlie's searching stare at that party in 1990, who knows how different things might have been? I could have spilt wine out of my glass and over my fingers, which is what usually happens when I'm nervous, said something crass, started eating crisps in an effort to sober up and defused the whole thing. Instead of which, I decided I couldn't cope right then and skipped off, and the look stayed there, emotionally undermining me. It was still burning through my posh frock, unfinished business, when we next met a year later. When Charlie and I finally did hold a conversation, it was underscored by the need to get over that awkwardness, by an awareness that our relationship was already on a whole other footing of emotional ambiguity and uncertain potential. (This is a not uncommon tactic of shy people, to spin fervid webs of feeling in the air around them, in an effort to transmit a sense of being fascinating without actually having to speak. If the relationship develops, it does so freighted with that moment of initial drama; if it doesn't, at least we haven't had to speak.)

So the little geyser of significance stayed there: a choice flunked beside the cheese and pineapple, perhaps because it appeared to be too momentous, and quite definitely set us up for something slightly different than if we'd known each other for ages and somehow, almost accidentally, drifted into a relationship.

For the house, on the other hand, there was no defining

moment, no hoarded, scary revelation. None of it seemed real. We drifted through that summer in a haze of 'Let's pretend' as Joyce and Ferhan progressively revealed plans for a house that might well never be built, partly because it was unimaginably unlike a house (an indoor/outdoor eating space: what was that?); but mainly because the planners had said they didn't want this kind of a building.

Every fortnight to three weeks, from the beginning of July to the middle of October, Charlie and I would drive from home to Islington, or meet on the corner of the alleyway where Joyce and Ferhan had their offices, climb the stairs to their eyrie and sit in the meeting space between the cupboard and the white wall, drinking coffee out of white cups served on a black tile. In my awe at the pristine control they exerted over their surroundings, this meeting area seemed ineffably sophisticated, although I now realize it was in fact a rather narrow space between a cupboard and a wall.

The mugs were always white, whereas at home we had mugs with pictures on them. One or two of them might even have been chipped. Presumably, if you had a Joyce and Ferhan house you were supposed to have Joyce and Ferhan mugs too, and this seemed more absurd to me than courtyard bathroom and glass walls. I had very little grasp of what these might be like. But mugs I understood, and we had a lot of crap ones. We were just playing at being people who had architecture.

But that in itself seemed worth the money. (Well, perhaps not quite the money we were spending, but *something*.) There's a lot to be said for playing at things – especially when you're an adult, especially, perhaps, when you're a female adult. I always like and warm to playfulness in other people – which is not quite the same thing as wit, though they often go together, or flirting, though flirting can be playfulness-with-intent. Not all children are playful, although they're more so, generally speaking, than adults. I like playfulness in myself, and in other people, and especially in people who bring it out in me, because it can be hard to access, what with getting older and having children and jobs. The ability to be playful seems to require generosity and a kind of self-forgetfulness, a readiness to take oneself not too seriously, which nevertheless remembers that people together can be more amusing than people alone; it's more a matter of demeanour than any specific activity. (Men in sheds are on the whole not playful.) Anyway, whatever it is, I felt we were getting a lot of it that summer. It was easy to feel that we were playing with Joyce and Ferhan, because they were always laughing, and because we were engaged in a kind of game. As Witold Rybczynski has pointed out, much of the satisfaction of architecture arises from 'the rules, the smallness, the fantasy of making space, the re-creation of childhood freedom'.

I spend most of my childhood pretending to be someone else, some kind of a princess usually, for the simple reason that frankly it was much easier being an Anglo-Saxon warrior queen or a girl-pirate-captain-cunningly-disguised-as-a-man than it was being me. There's a limit to how much you're allowed to do that as an adult. I spend quite a lot of time pretending to be a journalist; I go through the motions of being a mother, intermittently experiencing flashes of existential hilarity. The only time I can convincingly pretend to be someone else these days is when I'm ironing, and, for other reasons, I avoid doing a great deal of that. But being a client was a bit like being a cunningly disguised pirate queen: what if I were the sort of person who drifted around in Armani, drinking out of white mugs, what would my house look like then? Oh, like this, apparently. And it followed that if you had a house like that, how could you not be chic and poised?

Of my four children, the one who most resembles me in this respect is Freddie, also a child who seemed to feel in need of transformation, who was for ever dressing up or constructing elaborate Lego models in which good and bad were clearer and heroism came naturally. When Freddie became a teenager, when the dressing-up was no longer wholly appropriate and his younger brothers were only interested in having Lego demonstrations for a limited number of hours a week, he transferred his interest from little buildings made of plastic bricks to full-sized ones. He wanted to go to New York. I took him on a bonding trip when he was fifteen, expecting he would want to shop. But, mainly, he wanted to look up.

The word 'architect' derives from the Greek *architekton*, or head carpenter, while the Sanskrit and Chinese words for an architect – *sthapati* and *chientsu-shu* – both translate literally as 'master builder'. Frank Gehry, architect of the Bilbao Guggenheim, has said that his choice of career was influenced by spending much of his childhood playing with construction toys. The inventor of Lego, Ole Kirk Christensen, was a joiner, and Lincoln Logs, a sort of American Lego, were developed by John Lloyd Wright, son of the architect. Friedrich Froebel, arguably the inventor of custom-made, commercially available children's building blocks, began training as an architect before getting diverted into education, and introduced his blocks in 1837 as part of his Nine Gifts, a series of play materials of increasing complexity designed for use in the kindergarten, supplying a little song that children could sing as they piled his educative bricks on top of one another:

> A house, a house, a house!
> A house belongs to me.
> A house, a house, a house,
> Come here, come and see!

The lyrics could maybe do with some work, but the jaunty rhythms are about right for the little spritzes of excitement we felt as we sat across the table from Joyce and Ferhan and they turned over their big sheets of paper and we gazed at them, trying to read what they'd changed before they told us, to work out how much our lives were about to improve.

The use of A2 sheets of paper for the presentation of architectural plans is clever because, on A2, everything looks big. But not too big: the whole conception is simultaneously reduced to neatness and comprehensibility. Buildings appear either in plan – from overhead – or in section, in a slice from the side; and, unlike real-life structures, these formal shapes can't get out of hand. Not being three-dimensional, they don't acquire nuisance places at the back of cupboards containing cardboard boxes that you should really go through and chuck stuff out of. The drawings imposed order on the untidiness of our lives, on our family of trailing children, our toddlers and teenagers, who, from their point of view, half-belonged somewhere else, but from ours, wholly belonged with us. But more than merely imposing order, they *were* order. The process took on its own life; the game imposed its own rules. What we were doing at this stage with pencils and tracing paper had its own justification: the making of coherent plans. I couldn't envisage a finished building and it didn't bother me, because I had all these nice neat drawings.

Watching the building take shape on paper was rather like seeing fireworks go off: dazzled by the display, we were unable to focus on every streak of colour, every spark of light. We must have been frustrating: we'd fix on some passing twinkly thing we could grasp, but which was at this stage largely irrelevant, like whether we could have a bath in which we might sit up to read. And then we'd ignore much more fundamental questions, such as whether it was really a good idea to have Henrietta's bedroom next to the kitchen and garden, so creating potential for endless rows when the boys got up early and rode their tricycles along the terrace, right next to her bed, or even just ate their Rice Krispies noisily. I had a frequent sense that I was not only missing the wood for the trees, but also that I didn't know what kind of forested area I was looking for in the first place. I'd thought a lot about the interior design of existing spaces – how to make the through-room of a Victorian terrace look less like a corridor, where to put a sofa in a multi-purpose basement – but I'd never before had to consider making spaces out of nothing, and it was slightly brain-hurting, like thinking about that branch of physics that proposes parallel lives.

Ever since the concept meeting, I'd been walking around thinking happily of my spectacular hall with its glass walls giving on to vistas of bamboo and sky, which was more or less the only part of the plan I'd been able to get my head round properly. By the first proper design meeting, the hall and associated vistas had gone. The shape of the house had changed. Instead of a rectangle, it was now more of a Z-shape, although Joyce and Ferhan described it, in more elegant architecture-speak, as two interlocking cubes. It was as if someone had pushed the top end, by the lane, hard up against the mews, and the bottom end the other way, into the garden. You might think a moderately sentient client would ask why they had decided to do this. And perhaps we did.

But clients asking difficult questions are only a help to architects when they have yet to realize a mistake, rather than when they've already rectified it, and they must have given an answer so abstruse or distracting that we immediately forgot it again. They weren't about to start parading their errors, even though architecture is inevitably a process, and architects do spot better ways to do things all the time; sometimes they realize – or are told by the structural engineer – that something they've suggested would, in real life, fall down. But they like to retain an air of working-by-mystique rather than working-by-mistake. They prefer a pliant client, who doesn't probe the process too much, and I assume they learn how to turn you into one at architecture school. Joyce and Ferhan's favoured method was to flourish a revised plan, giving some vague justification for it ('We thought this would be better') and then setting off a lightning flare – in this case, that they would be moving the den out of the kitchen so that Harry could play his Nintendo 64 without directly interfering with *The Archers*. This was such an obvious improvement, requiring so much discussion, that it blinded us to the fact that the new plan created an area of garden in the north-east of the plot that would only ever be any good for slugs.

Later, when the existence of the slug garden perplexed me, I assumed it must have been imposed on us by the Planning Officer. In his initial discussions with Joyce and Ferhan, he must have said something about the need to respect the rear line of the mews, or that the Planning Department couldn't recommend approval of a glass wall that would allow us to look down on people sunbathing in the garden next door every time we climbed the stairs. (Next door was the back garden of something called the Foreign Missions Club, a guest-house for missionaries visiting London from foreign places. You might imagine that missionaries would have their minds on higher things than sunbathing, and also that they would get quite enough sun wherever their missions were, but that would be wrong.) In reality, I think, Joyce and Ferhan changed the floor plan because they simply couldn't get enough rooms into the one they'd first thought of. As for the hall, it might not now have a glass wall, on account of there being a den and a bathroom on the other side of it – but, they assured me, it could still be spectacular, especially if we left the sitting room at the other end of it open to the skylight.

'What, like not have a ceiling?' I asked.

'Not at one end of it. You'd have a void there. It would be very dramatic.'

Joyce and Ferhan's children were all at primary school. They had no idea what it was like to have teenagers who come in at midnight. So we explained to them about children reaching an age at which they talk and sleep at the wrong ends of the day, and they agreed to look at getting rid of the void over the sitting room. Then they explained the indoor/outdoor living space, which would be protected by a sliding metal grille, which seemed to me, not an especially hardy person, not very much protection at all. Still, no one else would have one. We would be marked out as radical and experimental people when we had friends and acquaintances round to dinner; and we could always warn them to bring jumpers.

I tried to describe the plans to my mother – at this stage we still weren't taking the A2 sheets out of Joyce and Ferhan's lovely meeting cupboard. 'What, so you're going to have a flat roof?' she asked doubtfully. 'Aren't you worried about leaking?' (Perhaps she had a point here: the first owner of Fallingwater, Frank Lloyd Wright's masterpiece, described it as a ten-bucket building.) Her other main piece of advice was to have somewhere by the front door to put coats and boots – prompted, I expect, by distaste for our current habit of accumulating coats over the end of the banister in a big pile until someone was coming round, when we would finally distribute them around the bedrooms. 'We must have somewhere to put coats,' I said to Joyce and Ferhan at the next meeting. 'We were always planning it,' they said. But the meeting after, it appeared. I was always sceptical about the studies on the roof, which stuck up like a kind of funnel, giving the house the appearance of an ocean liner, which I don't think was the effect Joyce and Ferhan were striving after at all. (Had they been postmodernists, a house that looked like an ocean liner would have been thrilling, but they weren't.) Besides, the studies effectively added an extra half-storey, which would mean either that our house would poke above the roofline proposed for the developer's houses or be required to have very low ceilings. It was already obvious that Freddie was going to be tall, and Henrietta's boyfriend Matthew was 6 foot 4: I envisaged grandchildren who thought of us as quaint figures out of history, like medievals, living behind doors that you had to stoop to get through.

It quickly became clear that building-control regulations meant that we'd need all sorts of ugly, heavy fire doors if we wanted these studies on the roof. So Joyce and Ferhan moved them elsewhere. 'Those studies on the roof never looked as though Joyce and Ferhan had considered them properly,' Charlie said to me thoughtfully once they'd disappeared, but – and this was typical of the way the process worked – not until then, because there were always too many other things to preoccupy us.

Charlie's study went out to the back of the garage, becoming a kind of shed at the bottom of the garden, although on the plans it was labelled 'second car space', because Joyce and Ferhan didn't want to alarm the planners by calling it 'office', thus giving the impression that Charlie was going to run some sort of car repair business out there, or an establishment involving the heavy use of drills (rather than, as in reality, the heavy use of pencils). My study was at the end of the upstairs hall, but didn't seem to have a door. The trouble was that even with the new interlocking cube arrangement, it was difficult to fit everything in. It was like one of those Christmas cracker puzzles with letters on a square of plastic that you have to push into the right order: you displaced one thing and it created problems somewhere else. The process was in this sense quite unlike a refurbishment, where you already have experience of inhabiting the space and there are certain known quantities. Not only was there a lot of information to absorb at every meeting (information that we were untutored to process) but the house looked so far away from any place I'd ever lived in that I didn't know how we *would* live in it, what preferences we could possibly have.

Le Corbusier once claimed that he sought, in his architecture, to exalt what he called the 'White World', which he explained as pure, streamlined, calming, with exact proportions and precise materials, over the 'Brown World' of clutter and compromise, which the art critic Robert Hughes sums up as 'the architecture of inattentive experience'. There was a part of me that believed, and certainly wanted to, that I was up to living in a White World, even though my world at present was exhaustively inattentive, indubitably brown.

For many reasons, I couldn't have embarked on this project without Charlie (money, most obviously, but also because, without him, I doubt I could ever have taken myself so seriously over such a long period of time). But he was also much more instinctively modernist than I was. When we first met, Charlie's house had, to my mind, shockingly little furniture. There was a single black sofa against the wall, a desk, and a low table supporting a small television. In this arrangement, his portable typewriter took on the appearance (quite appropriately, in fact) of a major item. There were bare boards everywhere and, for some reason that was incomprehensible to me, he had removed all the internal doors. The feng shui of this was, to me, terrible: you felt someone was going to walk in on you from round the corner at any moment. Which of course they weren't, because Charlie lived alone. But I wasn't used to that either. I was used to people barging in and dropping half a toy before wandering out again. I was used to a house littered with half-toys that I was too distracted, too plain exhausted, to pick up, let alone to hunt down their corresponding halves. Even my car was messy, with old drinks cartons dropped under the seats and books left behind from long journeys and sweets trodden into the carpeting. My house stood no chance.

I was less exhausted these days: there was something about living with Charlie that was energizing, so that I no longer looked at untidiness and felt hotly, furiously resentful and then defeated, on the verge of tears and overcome by the impossibility of imposing myself. Now I looked at mess more like normal people, and thought less hysterical, more normal thoughts, such as 'Oh fuck, better pick that up sometime.' But even living with Charlie hadn't made the children any less prone to leaving toys in the wrong room or kicking vital components of games under sofas. There was always cleaning to be done, tidying that could be done better. I was the daughter of a clever, just pre-Betty Friedan woman who found it hard to understand how her life had become about tidying up spaces only so that people who didn't take her seriously (i.e. us) could untidy them again. Torn between her love of order and her loathing of a role into which she had been forced, my mother retaliated in the only way open to her (since she loved my father): by suggesting to her daughters, never overtly but in a million tiny ways, that it was preferable to get a job and pay someone else to do your cleaning. That it was better not to care about it quite so much, not to think like she did.

Very recently, following some tests in hospital, my mother decided to spend a whole day pampering herself. So she wrote a long letter to a cousin in Australia and did her tax return. 'You will laugh at me,' she said, accurately, when she described this day of indulgence, 'but I couldn't think of anything nicer than sitting at a desk. I ignored the house and garden. For a whole day!' It struck me then how effective her childrearing had been, because I often lock on to my desk early in the morning and don't get up, except to make a cup of coffee, for hours. And also, how bad she is at allowing herself the leeway she wanted for us.

Unfortunately, to be the focused bluestocking my mother might have been had she not been born in the East End in the first third of the twentieth century, it is better not to care too much what your house looks like. I do still care. Gazza is said to suffer from a syndrome that renders him unable to leave the house until he's straightened all the towels. This sort of thing occasionally afflicts me, but I am much more troubled by a sort of mirror-image of it: I feel ferociously angry when I come home and the towels aren't straight, even though I've made no attempt – and I know this before I walk in – to straighten them. And for years this troubled me a lot because I employed cleaners who never cleaned very well, or certainly not as well as I believed I could have done if only I could have been bothered. (I also suffered from the unhelpful lack of clarity of the insecure employer in relation to cleaners: I was given to thinking of my grandmother, who had risen at five every morning to travel from her council flat in Stepney to work as an office cleaner in the City, even though these days she would have been registered blind. And my courage failed.)

So, alternately lured and threatened by Le Corbusier's White World, I looked up from Joyce and Ferhan's clean lines on A2 one afternoon in late July and asked, 'So what about the toys? Where do they all go?'

Ferhan frowned prettily. 'We both have a rule in our houses, Joyce and I,' she announced, the combination of her accent and her energy making any utterance sound like machine-gun fire. 'Only two toys out at any one time.'

'Only two toys?'

'Two big ones, obviously,' said Joyce, more emolliently.

I thought of our basement. The box full of Lego pieces. The other box full of Lego pieces. The box of bits of Playmobil castle. The box of miscellaneous toys: guns, balls, recorders. The old Bahraini cooking pot I still hoped would one day hold a plant, but which was actually full of soft toys. The box of rattles and assorted other baby toys: stacking hoops, shape sorters. The big red fire engine. The children's books on the windowsill. The latest half-made Lego model, also on the windowsill, which Freddie didn't want dismantled. Anyone would think they didn't have bedrooms. But they did, and those were full too.

And it wasn't as if I was undisciplined about the toys. Roughly one Saturday morning in three (I spent Saturday mornings doing all the bits of cleaning the cleaner had missed) I'd sort out the toys and throw away any that were irreparable. I had to do this when the children were out at their music class, because if I so much as inched towards a black polythene bag containing a bit of plastic in their presence, it immediately became the most important toy ever.

So it was inconceivable to me that we could only have two toys out at a time. And then my heart should have sunk, knowing I was incapable of living in a Le Corbusier-style White World. Fortunately, though, it was only a game.

We thought that a house that was designed around us and our needs was a radical proposition, but in reality it wasn't radical at all. The things we wanted were deeply predictable, conditioned as much by fashion as individuality. The requirements people have of houses change over time, but are pretty much universal at any given period. In Jane Austen's day, any self-respecting young woman was in need of a breakfast room, in case of marrying Mr Collins. In the early 1900s, a man might have wanted a smoking room (now available only in an aversion-therapy way in airports) and a billiard room, which disappeared in Britain but mutated in the US into the rec. room in the basement. Then people got more hygienic and wanted bigger bathrooms and more of them, ideally one off every bedroom. All these developments were driven by social imperatives: the American den, or television room, ceased to have a point once televisions spread into bedrooms. Central heating increased the likelihood of family members dispersing to different rooms, especially once they could access pornography on their laptops.

Right now, families are more different from one another and more fractured than they were a generation or two ago. They are also more self-conscious, resulting in pressures on the one hand for flexibility and on the other for places in the house where people can cohere and feel properly familial, for privacy and community. More people are working at home at least some of the time, and need offices. Children have quantities of toys. Aged parents are becoming more aged, and may need guest rooms, preferably on the ground floor. Childless couples and those with toddlers might like open spaces; parents of teenagers want rooms you can shut off. You may not be able to supply space for all of these things when you start from scratch, but at least you get more chance to think about them rationally.

But by saying I wanted the kitchen to be the heart of the house, I was being absolutely predictable and dull. I know of no one who has added an extension to their house in the last five years who hasn't wanted a wall of glass on to the garden, a seamless, indoor/outdoor space as the centre of domestic activity.

The Victorians, who had a horror of cooking smells, located their kitchens as far away from their living rooms as possible: in a terrace like ours in Hackney, that would have meant in the basement, but in some Victorian country houses the kitchen would have been 150 metres from the dining room, cold food evidently being preferable to the smell of boiled cabbage. Now we're back to more of a medieval hall arrangement, with everything happening in one room. A kitchen-designer friend of mine, Johnny Grey, says he no longer knows what to call the rooms on which he works, because people don't want just kitchen in their kitchens any more: they want a playroom, a television room, a place for the sofa, maybe a staircase, certainly somewhere to eat. The idea of sad people gravitating to the kitchen at parties doesn't make sense any more because there aren't really any other rooms.

It has also become more or less compulsory to have lots of glass in kitchens, so that we can pretend we're cooking in the garden. I don't know why this is (Delia Smith once did an entire cookery series – *Summer Cooking*, it must have been – with roses growing out of the back of her fridge, but I don't think it can have been that), unless it's true, as Le Corbusier said, that the history of architecture has been 'the struggle for the window' and, one way and another, the struggle has now been won. Flying buttresses and Gothic arches were such important architectural innovations because they freed walls to hold more and bigger windows. The mullions and transoms we now find so charming were introduced in response to a practical need to increase the size of the casement at a time when windowpanes were still made by blowing glass bubbles, flattening them out and cutting the biggest square possible from the resulting pancake. The introduction, in the nineteenth century, of sheet glass made with iron rollers changed everything, opening the way eventually (once structural steel made curtain walls possible) for a new conceptual language: it became possible to conceive of a building with a glazed skin, rather than merely glazed openings in a skeleton of brick or stone.

The property of glass of seeming to dematerialize boundaries must have seemed almost mystical to the modernists, fixated as they were by notions of transparency – whether of construction (so you could see how the building was put together), of function (no ornament for its own sake) or of space (not too many walls). Transparency implied truth and freedom, and glass, with its capacity to make walls seem incorporeal, seemed to offer political possibilities for breaking down the boundaries between human beings and their environment, between one group and another.

'To live in a glass house is a revolutionary value par excellence,' wrote the Marxist critic Walter Benjamin. 'It is also an intoxication, a moral exhibitionism, that we badly need.' Modernism shared Marxism's belief that progress is latent in technology, that the machine might help to free people from the slavery it had helped to create. And glass was the way in to Utopia. This filmy material could now clothe the side of buildings, a see-through skin allowing people to engage with the world beyond. Glass was almost an extension of our eyes, another sensitive and pervious membrane. It shone in the sunlight. In 1914, Paul Scheerbart, a German engineer and science fiction writer, proposed in a heady manifesto that the word *Fenster* might actually fall out of use, as windows gave way to walls of glass: 'The surface of the earth could change totally if brick buildings were replaced everywhere by glass architecture. It would be as if the Earth clothed itself in jewellery of brilliants and enamel. The splendour is absolutely unimaginable . . . and then we should have on earth more exquisite things than the gardens of the Arabian Nights. Then we would have a paradise on earth and would not need to gaze yearningly at the paradise in the sky.'

I wasn't thinking about any of this when I said I wanted a lot of glass – not even that it was kind of appropriate that Charlie should put his faith in glass given his intellectual background, although this thought did strike me as kind of cute later on. But maybe there is something in what the sociologist Richard Sennett says about the Enlightenment, that it 'conceived a person's life opened up to the environment as though one had flung a window open to fresh air'. And this, perhaps, was what we were after, sticking all those sliding doors on the back of our urban houses: openness, an ability to breathe in opportunity and possibility among the canyons and grimy backyards of the city.

When I first met Charlie, he had a day job as industrial editor of the *Financial Times* and a night job writing for *Marxism Today*, an obscure publication that was in thrall to Antonio Gramsci and preternaturally keen on an (also obscure) politician called Tony Blair. He (Charlie, not Tony, disappointingly) recently said over dinner that he would still describe himself as a Marxist. When I queried this, he said he meant in the sense of believing that an understanding of the means of production is crucial to understanding the social relations of production. (He doesn't actually talk like this most of the time. In fact, getting him to talk about his ideas at all is a struggle because he has a horror of sounding as if he's still in the junior common room.) *Living on Thin Air* was clearly Marxist in this sense, in that it was about how politics and social organization needed to shift to take account of new technological realities; and the book he wrote while we were building the house was a protest against knee-jerk anti-globalization, an attack on the pervasive pessimism of society in the face of change.

But, as I say, I wasn't thinking, 'How sweet and appropriate that Charlie and the house will share theoretical antecedents to do with Marxism and optimism' but, 'I want one of those walls of glass like everyone else.' I, too, wanted to be able to throw my doors open in summer and have the garden come into the house, or the house go into the garden. According to the garden designer Stephen Woodhams, who admittedly has a vested interest, this vogue for incorporating the garden rather than viewing it as an add-on backyard means it is in the process of overtaking the kitchen and bathroom as the 'room' people obsess about most and expect to clinch the sale of their property. But I think it's slightly more complicated: the kitchen and the garden have merged, and both have to look not only right, but also complementary (preferably with some natty inside/outside motif: the same flooring, or a wall that just carries on going when it hits the open air); they have to give every appearance of being part of a seamless conception. And then, of course, you have to avoid letting the garden part of it die.

I had actually lived in a building with a wall of glass once before: in Bahrain, where, for a while, I had a long, open-plan, one-storey house on a compound (which is an expatriate word for housing estate), with sliding windows on to a terrace facing the street. This architecture, in that place, was made possible only by air-conditioning, and it seemed typical of the flagrant extravagance of the place generally that in temperatures nudging 50°C, anyone could think that a lot of glass was a good idea. The indigenous architecture featured thick walls, heavy wooden doors, courtyards with high sides and wind towers to circulate the air. My house was a defiant slice of American suburbia plonked down in the desert, wasteful of resources, heedless of its surroundings, showing off its affluence. It seemed absurd and wasteful and inappropriate, and I liked it.

If glass can meet some visceral desire to merge with the light in the baking desert, then certainly it must in London's drizzled streets, under overcast skies. Rather as we in Britain buy more convertible cars than any other nation in Europe, we seem to have recognized that we owe it to ourselves to have glass walls.

By August, the house was more or less imagined. You'd come through a front door in a nearly blank wall facing on to the lane into a hall that was open to a rooflight above. This would run the length of the building until you reached the sitting room, transverse across the other end. To your left would be, first, Henrietta's bedroom, then the large kitchen and dining space, all with walls of glass on to the garden. To your right, there'd be a bathroom and the den.

Upstairs, there was another long hall. The master bedroom would be above the sitting room, again running all the way across the bottom, northern, end of the building, incorporating a dressing room and bathroom, plus a balcony off to the side, above the indoor/outdoor eating space. Ned's, Harry's and Freddie's bedrooms would sit along the western side of the house, overlooking the garden, and at the south-eastern end, in the other side of the house (the area created by the cube), there was a laundry/store room and a bathroom.

We didn't need to know at this stage which way round the kitchen units would go, or whether our bedroom space would be configured dressing area/bathroom/bedroom, or bathroom/dressing area/bedroom. But we did have to be sure about the number of rooms and their sizes, because we wouldn't be able to change anything about the basic structure or configuration if and when planning permission was granted, other than by going through the whole process again, which would piss off the Planning Department and Committee, and also us.

We took the plans home and showed them to people who screwed up their eyes and said 'Hmm' with varying degrees of mystification. Usually we pointed a couple of things out, to help them. At the outset, Charlie had suggested we should each specify something for the house. I nominated the kitchen with glass walls, Henrietta and Freddie wanted telephones and laptops in their rooms, Harry, a tree-house, and Charlie, a place to shower outside. Early on we'd talked about an internal, open-air upstairs courtyard, incorporated into our bedroom area, but the pressures on space were too great (not least since I had a disinclination to shower outside other than on about three days a year), so this had mutated into a bath with a retractable rooflight above, meaning it would be possible to bath and shower in the open air, but not compulsory. When people asked us what the house would be like, we'd say airily, 'Oh, it'll have a bath you can sit up in to read, with a glass roof above it that opens to the sky. And there'll be an indoor/outdoor eating space.'

Hugo, I could tell, was looking at the plans in a quite different way from most people. I was used to seeing people's eyes darting around in panic, trying to fix on a starting point; Hugo's moved methodically across the paper, tracing the lines of traffic in the house, working out the position of the sun at different times of day, assessing the light into the rooms; translating, in an alchemy of imagination, the marks on the paper into three-dimensional spaces.

'You want to be careful about this indoor/outdoor eating space,' he said. 'Make sure it's big enough for a table.'

'What, you think it isn't?'

'Well, you could get a table there, for sure, but not a very big one. You're probably only talking about four people.'

'So you'd definitely need another dining table?'

'Definitely. And then mightn't they look funny together?'

'I suppose they might. They'd sort of clash, wouldn't they?'

By the next meeting, in September, the indoor/outdoor eating space had gone.

'I spoke to Joyce about that space for your table,' Hugo said one morning on the school run. 'She said they were already thinking about it.'

I don't know what finally did for the indoor/outdoor eating space. Joyce and Ferhan later claimed it would have been too expensive. Brian Eckersley, the structural engineer, whom we had not yet officially appointed but had agreed informally should work on the project (because Joyce and Ferhan said he should), once suggested that the house would have fallen down if it had remained – or maybe that was the rooflight over the bench in the kitchen, which also went. But the fundamental problem was that it suffered from a serious drawback as a space designed for dining: it wasn't big enough for a table.

At the September meeting we also saw a series of computer-generated, swimmy grey-green three-dimensional pictures, glaucously unreal, at least one of them taken from an angle at which no one would ever see the house unless they were capable of levitating 20 feet from the lane. These pictures excited but perplexed me: was this what we'd been designing? I was pleased to see how little of the building would face on to the lane, and how much would be hidden behind a high brick wall; I was encouraged by the expanse of glass, and by how strikingly modern the house-animation looked, but I also felt alienated from it, even slightly repelled. The house looked flimsy in these pictures. It lacked the solidity and richness of real houses, even of photographs of real houses; it had the flat, opaque quality of computer-generation, a strange, in-yer-face *Toy Story* feel, splat, there in front of you. It hid nothing, and had no way of inviting you in because it held no secrets, its interiors just the back of a computer.

By October 2000, Joyce and Ferhan were presenting the drawings they wanted to send to the Planning Department. These showed shutters on some of the windows (from the roof to the ground at three points along the garden side of the building) and included elevations that suggested that the walls at either end, including the one on to the lane, were to be made of pre-cast concrete panels, with a big timber doorway at the front. I imagined these walls rendered a pristine white, symbol of my new White World.

By 23 October Joyce and Ferhan were ready to despatch a bundle of computer-generated images, site photos and investigations, plans, elevations and an application for conservation area consent to Islington Council's Planning Department. We all congratulated ourselves on not having done badly: it had taken a little under four months – including everyone's summer holidays – to design the outline of the house.

We received a letter from the Department acknowledging our application and informing us that we should receive a decision by 21 December. If we hadn't, we could appeal to the Secretary of State for the Environment in accordance with sections 78 and 79 of the Town and Country Planning Act.

A few weeks before Christmas, Charlie and I went to a party thrown by Atlas Venture, dinner in a private room at the Hempel Hotel. It was our second Atlas party of the year: the first had been at the Oxfordshire home of one of the senior partners, a manor house of honey-coloured stone surrounded by water. The sun had shone as we strolled in the garden, played tennis and eaten strawberries, and the sunshiny, movie-backdrop mood seemed buoyed on a tide of money that it was hard to believe would ever stop rolling for these people, even though, at that point, the Nasdaq had already collapsed.

By Christmas, Atlas was restructuring. There were at least three people at dinner who were on the point of leaving, some to return to an academic life cushioned by affluence, others with barely concealed bitterness. As we made our way back east at the end of the evening, Charlie and I wondered whether we'd be there the following year.

I can't say the anxiety troubled me as much as it perhaps should have done. I found the people at Atlas urbane, charming, witty, clever, and mad. It was difficult getting my head around their wealth, but more especially around their interest in wealth. They were delightful, except that they thought about money *all the time*. And the money-madness seemed to me like a growth on the soul: it distorted everything else.

December 21 passed without a whisper from the Planning Department.

I called Joyce. 'What about this appeal? Are we supposed to activate it?'

She laughed. 'Of course not. If you did, it would take a year for the case to be heard, and you still wouldn't have a decision. It's just a formality. No one ever does it.'

Things took a while to gear up again in January since, increasingly, Britain seems to lose about six weeks' work in the winter, which is fine in theory, is something I would support, in theory, being in tune with the rhythms of the year and the natural desire to bed down through the winter solstice, but is bloody annoying when you are losing £87 a day.

By February, things had got going again sufficiently for us to be allotted a new Planning Officer and for him to ask for a change to the drawings. The front door was to have been concealed in a little indentation or cubbyhole on the front wall, so that you would effectively step into a porch and turn left to enter the building. Mr Armsby was concerned that muggers could lurk here and jump out on passers-by, or vagrants take up residence. So we put the front door on the face of the wall (we might get wet while we fumbled for our keys, but at least we wouldn't be tripping over tramps). For similar security reasons, we were also asked to add a couple of windows to the front of the building. And the 'two-car garage' had to be replaced with a one-car garage. A somewhat truncated space for Charlie's office was now labelled 'garden store'. And then, once we'd done all this, we were finally given a date for a hearing in front of the Planning Committee: 27 March. By then we would have spent £21,141 on the bridging loan.

5

Islington Town Hall is plonked down heedlessly amongst the jostle of bars and boutiques along Upper Street: a slug of white classicism displaying too much frontage, it leers over the long stretch of pavement that it has somehow managed to turn into wasteland, its forecourt littered with green and white corporate identity signs designed to communicate user-friendliness but somehow redolent of underfunding. The logo-loud placards sit awkwardly against self-important pillars and pediments, carved to proclaim that the building was completed in MCMXXV.

The Town Hall is made of stern stuff, constructed at a time when it was necessary to hang on to certainty. There are more steps inside, marble this time, with brass handrails, taking you up steeply to the heavy wooden doors of the Council Chamber, all the grave flummery of the materials designed to make the point that local government matters – which of course to us, on the evening of Tuesday, 27 March 2001, it very much did.

Ferhan, who was to speak, was ushered to a forward seat in the tiered Council Chamber, while Joyce, Charlie and I were shunted to somewhere near the back. Joyce and Ferhan had explained the format. Around thirty schemes would be discussed that night. Each project would be announced and the case officer responsible for it – the local government official – identified. The Planning Committee, elected politicians who would already have visited the site and received a recommendation from the Planning Officer, would then ask whether there were any objections, which would be registered with a show of hands. The objectors had to nominate someone to sum up their reservations; this person would be allowed to address the meeting for a maximum of five minutes. Then someone from our side could respond, again for five minutes. Since Joyce was managing our project, she should have done it, but in fact Ferhan would speak for us because Joyce believed Mr Armsby didn't like her (something to do with a dispute over the house that she and Bill had built for themselves; I never really got to the bottom of this, because the discussion would always get sidetracked into what Ferhan should wear. She favoured low-cut).

I scanned the agenda. We were item twenty-three of thirty. By item three I was already so tense I could barely swallow. The Committee swept through the construction of small flats in the backyards of pubs, attic conversions, the change of use from shops to minicab offices, school extensions. An eight-storey bank building on the corner of Finsbury Square on the edge of the City took a bit longer, owing to the existence of some architects' models and a sense of civic importance. An attempt by a builders' merchants to turn a green space up the Holloway Road into a car park had brought out many softly-spoken, liberal objectors, who glared, in a hurt manner, at the large contingent from the shop. The full complement of builders' merchant staff seemed to have turned out, along with most of their friends. Louder than the protestors, and better pleased, to judge from their dress, with their bodies, they competed with their opponents' undercurrent of 'tsk' and 'peugh' noises with a chorus of 'Shit!' 'Crap!' and 'Bollocks!'

So it was quite interesting, or would have been if my head hadn't felt as if it were floating 3 feet above my body. A lot of people cleared out after the park was saved, but, by item twenty, it was obvious that there were still far too many people left in the Chamber to account for the handful of cases that were left.

'I hope,' I whispered to Charlie, 'these aren't all here for us.' He was looking pale, and told me afterwards that it had simply never occurred to him until he got to the meeting that there might be objectors, nor even, really, that the Committee could say no.

A man sitting not far away from Ferhan seemed to me particularly worrying. He looked like a lawyer, to judge from his grey suit, which was designed to draw as little attention to its wearer as possible, and the way he was clutching a large sheaf of papers (sort of professionally). That was all we needed, I thought dizzily – some brilliant advocate applying fierce forensic intelligence to our little project, inserting stilettos of reason into the soft underbelly of our fantasies.

Eventually, after an interminable couple of hours of feeling as though I'd drunk about twenty cups of coffee and was buzzing all over, they called our number. And, sure enough, hands went up all over the hall, including from lawyer-man. He was their spokesman, and he launched in quickly – which was vital, because, though concentration was difficult with my head so far removed from the rest of my blood supply, I was conscious that he had an awful lot to say. There was something about the fate of the ash tree, and the noise from our washing machine in the flat in the next-door mews, and the fact that our garage would be close to his house and might damage his property. He said the Planning Committee had not taken into account a window of his that would be overlooked by our building. Plus, he got his car out by reversing down the lane; he was worried that we would do the same and collide . . . and on he went, and on, for so long that the Chair had to hurry him up.

Ferhan responded in her crisp manner. She explained exactly how the tree would be protected during the building work, and about the sound insulation. She said the party-wall issues would be dealt with through the statutory procedures, that the window to which he referred was small and high, and that when he came out of his garage, she would suggest – well, looking. (With Ferhan, you can't always tell whether she means to be rude, or is actually being perfectly polite, but in a Turkish way.)

The Chair asked for the Committee's comments.

'I have never,' said one, 'heard so many petty, trivial objections – and to such an excellent scheme.'

The Chair nodded. 'This promises to be a house of outstanding architectural interest,' he said, 'If I'd known about the site, I would have tried to buy it myself. Really, my only regret about this is that it will be hidden down a little lane and very few people will see it.'

It was scarcely believable. They *loved* it. They passed it unanimously.

Outside, I threw my arms around Joyce and Ferhan. 'Hugo will be *so* jealous,' Joyce said happily. We were staggering down the steps, Ferhan congratulating herself on having chosen the right outfit, everyone else congratulating her on having been so calm and smart, when the lawyer introduced himself as our new next-door neighbour. I felt sorry for him: we were all grinning stupidly, like drunks unable to sober up, and he'd just been accused of being petty and trivial in a public place. It was decent of him, and I did, honestly, struggle to stop smiling while we exchanged embarrassed English pleasantries.

We wandered down Upper Street looking for somewhere to eat, since we were all too excited to go home and suddenly realized we were hungry. Nowhere seemed quite special enough until we reached Granita, which was famously special as the location of a crucial, power-trading, top-level Blair–Brown meeting: it seemed appropriate, if we were to become residents of Islington, that we should eat in a restaurant so Islington that it was a cliché. (Not long before, when Charlie had told Tony Blair that he lived in London Fields, which was where the Blairs had had their first house together, Blair had responded, 'What, *still*?' as if, because he now lived at Chequers, we should too.)

Tony Blair wasn't in tonight; but there, at a table by the window, were Hugo and Sue, which couldn't have been neater if we'd organized it. Hugo was indeed delightfully, gratifying, jealous, but decent enough about it to advise us on the wine.

★

I hadn't been to the land for months. The last time, as far as I could remember, had been on Boxing Day, in an attempt to prevent Harry and Ned from noticing that lunch was an hour later than usual. The boys couldn't have been less interested in trailing about for a chilly half-hour on a bleak piece of land on which nothing was happening, where only the weeds seemed to be having any fun. The weeds were presumably doing even better now that the ground had warmed up, but I hadn't been to see because once the plans were lodged with Islington Council, it felt like masochism, envisaging a house that I might not be allowed to have.

But now that we were really and truly going to build, I wanted to imagine myself living on that little piece of land in a house with a retractable rooflight over the bath and a glass wall on to the garden, to picture myself as a woman with white mugs and Armani.

When I drove down the lane on that Wednesday morning, the first troubling thing I noticed was a small gate in the wall to the right, bordering the factory/workshops, and beside it, a sign. 'Rose Cottage,' it said improbably, '1 Ivy Grove Lane.' It didn't look as though anyone used this gate, which seemed to be painted shut. It was a ridiculous name, wholly inappropriate for a bit of factory. All the same, whoever worked there quite clearly had our address.

The second, still more troubling thing was that the lane appeared to have got shorter. Roads do not, as a rule, do this, but here there was a good reason for it: a pair of large, metal, electronic gates had been erected some way up it. When I say some way up, I mean too far. Beyond these gates, The Glassworks had been chi-chi'd up into mews houses with little balconies and fashionably industrial metal windows. The once scruffy muddy track down to the dead-end had been gravelled over and landscaped with clumps of bamboo, silver birch and alchemilla mollis.

This was all perfectly nice, in its place, but its place was currently blocking off the roadway about one-third of the way up the side of our property.

Even this might not have mattered so much, but this one-third of our property was where our house was meant to go. In other words, our projected front door, our *approved* front door, was beyond a pair of electronic metal gates, which, since they did not belong to us, we had no means of opening.

I went home and called the estate agent for the name of the developer. In the stately dance that is employment in estate agencies, Sue Reynolds had moved on to a rival firm across the road. But one of her former colleagues seemed to know all about the mews. She thought the developer was a Mr Christalides 'or something like that' of Dome Developments in St Peter's Street. There was no Dome Developments listed by Directory Inquiries in St Peter's Street or, indeed, in London. There was no Christalides or something like that listed in the Business phone book. There were plenty of possibilities, similarish names, in the domestic directory (Christodalou? Christodoulides? Christodoulopolous?) but it could take weeks to get through them, what with having to work up the pronunciation.

That evening, I insisted that Charlie come with me to the lane to inspect the gates, as if I might, myself, have been mistaken. But, no: the access to our property, our putative modernist dream, had incontrovertibly been blocked off.

Who did this person with the complicated name beginning with C think he was? The lane was unadopted. Everyone knew that. Equally, everyone knew that its ownership was a mystery. We had a letter from Islington Council saying so. It was dated 15 June 2000, and signed by W. Barrows, Assistant Group Planning Officer. We'd needed it for the indemnity insurance. 'You say that your architects have inspected the property file and found no evidence of a claim of ownership to the lane,' Mr W. Barrows had written. 'I have no reason to doubt this; my clear understanding from more than fifteen years of dealings involving the lane – and there had been a number of high-profile schemes – is that its ownership is not known.'

The developer was not, in fact, a person beginning with C. His name was Tom Tasou. I eventually found out because some of his workmen were finishing off on the site a couple of days later when I was making one of my increasingly frequent and distracted visits (unusually, on this occasion, not merely of histrionic, but of practical value). The workmen told me that Tom Tasou did in fact trade as Dome Developments. But – and this was somehow typical of the man, of his not needing to do things the dully bureaucratic way – he didn't see the need to put this name in the phone book. There was however, a listing for Tasou Associates, and the address was in Islington.

'Ah,' he said, when I got through, 'I've been expecting your call.'

We agreed to meet at the site the following day. I called Azman Owens, since this looked like a job for Joyce (everything, these days, looked like a job for Joyce), and got through to Ferhan. 'He must have put those gates up after the Planning Committee did their site visit,' I wailed. 'The planners *hate* gated developments. They were vitriolic about them at our meeting.' (Two schemes had come up: it was very clear that Islington councillors had a strong political animus against rich people shutting out the rest of the borough's residents.) 'And they'd have seen that our plans were unworkable with these gates. Surely Mr Armsby would have said . . .'

'He will have a price,' Ferhan said briskly. 'Everybody has a price.'

Did they? I thought about this, and wondered if I did.

The following morning, Joyce, Charlie and I waited nearly half an hour for Tom Tasou. When he eventually turned up, he parked his car, got out, went and stood proprietorially beside his gates and waited for us to speak. He was a small man, and something about him made me think that he had more energy for this than I did.

'Look,' I began uneasily, 'since we bought our land and designed our house these gates have gone up. I don't believe you can have planning permission for them.'

(I don't know why I was doing the talking; Charlie is much better than I am at not getting tense. And when he loses his temper it looks much less silly.)

'It doesn't matter whether I have planning permission or not,' Tom Tasou said smoothly. 'This is my land, and if I say you can't cross it, you can't.'

'But how can it be your land?' I said. 'The lane's unadopted. No one knows who owns it.'

'If you don't believe me, look it up in the Land Registry.'

'I see,' I lied. 'Well, we have a problem, because we've just got planning permission for this house' – I unfurled the plans – 'and the front door is here.' I gestured beyond the gates.

'Hmm.' He looked at the plans, feigning interest; it was only later that I realized that of course he had looked at them in quite some detail already.

'So,' he said at length, 'I have something you want. We have to come to some arrangement. You understand my meaning?'

This was a favourite phrase of his. He said 'You understand my meaning?' about ten times more in the course of this conversation, like a nervous tic, presumably imagining that he moved too fast, mentally, for anyone else to keep up. It didn't particularly bother me, perhaps because at some level I really didn't want to understand what he was saying, and he was right to wonder whether I was following – but it drove Charlie crazy.

'This is my land,' he repeated. 'You understand my meaning? You want to talk to me.'

We didn't, actually. We wanted to kill him. 'You mean you would be prepared to sell us the land?' I asked wearily.

'I'm open to listening to what you've got to say to me.' He was like a cat with a couple of dead mice. 'And that's a possibility.'

'So, er, what other possibilities are there?'

'You could acquire the right to cross it. It amounts to the same thing. And it doesn't bother me.'

We agreed to go home to think about it.

'Nice house,' Tom Tasou said, as I rolled up the plans.

I drove home, seething, and punched our lawyer's number into the phone violently. 'How can this be,' I wailed when I finally got through, 'when we all thought that no one knew who owned the lane?'

He said something about having pointed out that an area of the lane around the factory was in fact owned by the factory.

'But not up to and past our front door!'

'You didn't know where your front door would be then.'

As he was speaking, I did vaguely recall his having asked us whether we'd need to take the car down to the bottom of the site. But in the flurry of activity I hadn't registered the significance of the question. And, definitely, it had only ever concerned the car. Had he queried whether I might want to walk down the lane the full length of my own property, I might have asked what on earth he was going on about.

Clearly, it was his fault. He should have pointed out that if we put our house at the farther end of the site, we wouldn't be able to reach it. He should have insisted that, never mind the rush, this was an important point. He should have issued some kind of written warning. The trouble was, I knew it was also my fault, for insisting everything should be done in a hurry, for not using a qualified solicitor, for not listening very, very carefully. I was probably thinking, when he mentioned it, of newspaper deadlines and whether Harry had clean PE kit and how many people would be in for supper and to what extent I had fulfilled my lifetime's ambitions and whether I still fitted into my black trousers. In my defence, though, the very faint, patchy recollection of his saying anything at all definitely included his not being very bothered about it.

'So what about the indemnity insurance?' I asked wearily. 'This is what it's for, presumably?'

'I'll have a look,' he said cautiously, 'but I don't think so. I think it's in case something should turn up after the event. We knew about this.'

Except, of course, we didn't.

A couple of days later, Charlie called Tom Tasou and offered him £15,000.

'You're in the right region,' he replied. 'But we need to talk about the price.'

We got his meaning. A couple of days later, we called back and offered £17,500. He said he'd think about it.

Then he rang us and said it wasn't really worth his while for this amount of money. But not to panic: we were definitely in the right region.

'We can't afford twenty thousand,' Charlie said helplessly.

'That's a pity,' said Tom Tasou.

Charlie had to bear the brunt of these negotiations, because I selfishly declared that I didn't want to deal with him. I don't think I lack energy, generally speaking, but Tom Tasou made me feel tired. He had too much relish for the contest. Negotiating with him, unless you had a similar verve for the thing, could never be other than a dispiriting experience. It seemed to me that to get involved, to cultivate the emotional engagement, threatened to bring on self-hatred.

We caved in and offered him the £20,000. It could have been worse, I kept saying to Charlie: he had us over a barrel; he could have thought of a number then doubled it. He could have demanded £30,000, or £50,000. In fact, of course, he'd demanded precisely what he'd gauged we could stand without being provoked into a redesign which would reorientate the house, or into selling the land. He'd asked us to think of a number and he'd added some.

He agreed to the £20,000. But then he announced he only wanted to grant us a right to cross the land, whereas we'd thought we were negotiating a right to buy.

By this point, I'd decided we needed a new solicitor, one with qualifications this time. I asked a lawyer friend of mine to recommend someone. Bridget, our new solicitor, explained that what Tom Tasou probably had in mind was an easement, which would be appended to the deeds of our land, and would give anyone who owned our property rights to cross the lane in perpetuity. He was right that it wasn't really any different from owning the land in its effect, so it was a bit of a mystery why he didn't want to sell.

We consoled ourselves with the thought that, having spent this money, we would effectively have a spare place to put a car, across the lane. Even if it would be the most expensive parking space in London.

Tom Tasou required us to fund the moving of the gates and their electronic reinstallation. We also undertook to pay for the construction of a bin store (at this point his tenants were leaving their bin bags outside what would one day be our house) and to relocate all their letterboxes and entryphones. And all inside six months.

When finally we were ready to draw up documents, Tom Tasou said, oh, no, he'd just realized, he actually wanted £25,000. There was some nonsense story attached (which made it, if anything, even more humiliating): some relative of his – brother, brother-in-law – now owned one of the houses and he needed compensating for his 'loss' too. At this point I would have spent a substantial amount of money never to have to utter the words 'Tom Tasou' again.

And, he added, he'd take £10,000 in the contract, and the rest in cash.

I never did find out how Tom Tasou owned that chunk of the lane (although once he'd told me to check the Land Registry, I never doubted he did own it). One of the neighbours, someone who'd lived around the lane for a long time, whispered darkly that a line had somehow been drawn in the wrong place. Obviously the thing to do was find out who'd sold to him, so I wrote to the Land Registry and encountered the same old problems. The Land Registry is bound only to disclose current ownership. To persuade them to tell you anything about previous owners, you have to prove you're involved in a dispute. Our spat with Tom Tasou, apparently, didn't count. (God knows what a real dispute would feel like.)

Presumably the Land Registry people are worried that if they were to give out too much information, it might foment revolution, inflammatory politics, etc. (although not being allowed to know who formerly owned your own land seems pretty politically inflammatory to me). I pottered nerdishly in the London Metropolitan Archives instead, and managed to discover that in 1829, our land and all the land around had been fields. By 1862 there were big houses behind, and our road appears to have been called Ivy Grove Mews. There were no buildings on our side of the lane then, although, ten years later, on the first edition of the Ordnance Survey Map, there was something on our plot, plus other structures beyond us down the lane. It's difficult to tell, but it looks as though there might have been some sort of little gatehouse on our site, which would give weight to the notion that it was always part of the same package of land as the buildings that became The Glassworks.

The factory across the lane was built on former market gardens in 1928, for a company called A. C. Cosser, which made broadcasting equipment. As the Second World War approached, they carried out much of the crucial research into radar in Rose Cottage (which, according to the company's official history, then had 'an almost country aspect, with the adjacent garden of old-world flowers, beehives and tennis court'). Cosser constructed all the receivers for the chain home radar system that eventually covered the country and by the end of the war, almost every combat aircraft operated by the Royal Air Force carried at least one item of Cosser-made equipment.

Not surprisingly, the Nazis repeatedly tried to bomb the factory. But they failed, and instead hit many of the buildings around it. On the (wonderfully comprehensive, war effort-style) maps of bomb damage issued by the London County Council after the war, the house behind ours is graded dark red, i.e. 'seriously damaged, doubtful if repairable', while the building on our site is bright red, i.e. 'seriously damaged, but repairable at cost'.

The Victorian house in the avenue behind us was pulled down to make way for the present villa in the early 1950s, but clearly the cost of repairing 'our' mews was too great for such a scrotty little building and it was gone, the land absorbed into the villa's garden, now owned by the Robertos, ice-cream merchants and one of the numerous Italian families in the area. (Even today, there are three Italian delis in the little parade of shops up the road.)

The mews next door to ours was designated a 'ruin' in 1954. But by 1968, an Ordnance Survey map shows it rebuilt, while Tom Tasou's mews houses and the land outside them (but interestingly, not beyond) are marked as a separate property called 'The Works'.

The villa and our plot was sold in 1997 to a Mr Eliades (possibly not unrelated to Mr Christalides), who planned to knock down the house, replace it with flats, and build the two houses on our site. His architect, Stanley Haynes, told me that neither he nor his client had been aware of any problem with access to the houses. On the other hand, Mr Eliades did decide not to proceed; I'd assumed this was because the fierce residents who ran the private road behind had raised so many planning objections and held him up for so long that by the time he finally got permission his finances were weakened. But maybe there were other reasons.

Clear as mud, then. Tom Tasou could have solved the mystery, but I wasn't going to ask him. I do know that the deed of easement names the owner of the land as Dome Developments Ltd; a few months later the land was sold to Firstpaint Ltd for an undisclosed sum. Firstpaint Ltd shares an address with Tom Tasou's other businesses.

I also know that when I was visiting the land one day I met a couple more of Tom Tasou's workmen who were putting finishing touches to the rental properties. 'Oh yeah, these gates,' one of them said to me. 'He told us not to take too much trouble putting them up, because they'd only have to come down again.'

And down they came, although not until one morning, about a month after we'd got planning permission, when Charlie went to Barclays Bank in Holborn and collected £15,000 in £50 notes. He loaded bundle after bundle into a Nike shoe bag, put the bag on his front, strapped over his shoulders, and buttoned up his coat over the top. Then he took a taxi up to the office of Dome Developments, where Tom Tasou and his brother made him sit and wait and watch while they counted every last note.

That February, Charlie left Atlas. Unlike a lot of people caught up in the dotcom bubble, he hadn't been paid in slips of paper that promised miraculous wealth a few months down the line; he'd had money. But now, as for some time it had been apparent it would, the money was drying up.

It wasn't a particularly good time for this to happen. We'd increased the mortgage on our Hackney house and were paying a whole other mortgage for the land. We still had the bridging loan. The construction would be paid for entirely out of borrowings, but the architects' fees had to be found from current income. And we'd just given every last penny we had to Tom Tasou.

Our very first meeting with Steve Symonds had involved one of his many useful mates, this one from Barclays' mortgage department, and the whole deal had been set up to be seamless. We'd get the bridging loan to buy the land, and then, once we moved on site, we would be able to get a self-build mortgage. This is the commonest way that people raise money to build a house, and allows for money to be released in a number of pre-arranged stages, usually between five and seven, so that you can pay for materials and the builders as the project progresses. Once the build-out is complete, the outstanding debt becomes an ordinary mortgage (in our case a large one, consolidating the mortgage on the land and the construction costs).

Unfortunately, in the time it had taken us to get planning permission, Barclays had bought the Woolwich and integrated its mortgage services with theirs. Not least because we'd been so dilatory about getting medicals, it was no longer possible to do the original deal. We had to go through the whole application process again. I got rather fretful over this, not really believing anyone would be mad enough to lend us all this money, and kept wondering aloud and irritatingly whether Steve would have the same pull with these new people. Would the former Woolwich staff use our deal as an opportunity to assert their political independence of people like Steve who made decisions by himself rather than according to the rules? (The rules, obviously, would never let us anywhere near such a large sum.)

Charlie spoke to Steve more often than I did and was consequently less neurotic, but when he had to prepare another set of accounts, he certainly – as he put it, Steve-like – 'wasn't in the business of minimizing' his income. Until then, he'd been paying rather low levels of tax based on his early freelance earnings, which were minimal in comparison with those of recent months, although in reality respectable enough, being around the size of mine. Our accountant warned him that there would be a very large tax bill coming.

The Atlas monies had meanwhile pleasantly removed the need for Charlie to hustle for work, and he was out of the habit. The previous summer, when he might have picked up projects that would have been paying now, he'd wasted ages working with a politician on a book that didn't, in the end, come off. Meanwhile, the dotcom collapse was feeding through into the rest of the economy, or certainly the bits of it in which Charlie was engaged. He had a contract to advise an organization that was helping local authorities to acquire appropriate technology; in April, they simply stopped paying. And he was doing some sort of strategic thinking for Channel 4, where advertising revenues had slumped: the payments came later and later and later, and required more and more chasing. Eventually, they cut his fees in half.

In the middle of May, having finally got planning permission and access to our land, we could get back to meetings with Joyce and Ferhan. The next phase would involve detailing all the rooms, deciding on materials, organization, cupboards, lighting, where the fridge would go . . . but first they wanted a business meeting, the purpose of which was to establish exactly how much we had to spend. The house was planned at 1,900 square feet, excluding the garage, for which they were setting aside £25,000, and the hard landscaping of the garden (the terrace, the high wall along the lane), for which they estimated we'd need between £15,000 and £27,000. (Up to this point, I thought we were designing a house of about 2,500 square feet. But they seemed to be fitting everything in. And since I had no idea how big 1,900 square feet was, it was difficult to find any reason to object. Besides which, it was too late.)

When we'd originally talked about whether we could afford to build a house, they'd said something about needing to think in terms of £150 per square foot. Now they explained that there were three possible levels of finish, which we could think of as costing £170, £200 or £220 per square foot. They showed us magazine pictures of a house that had been finished at the £220 level. Typically, I had no idea what I was looking for, and admired the way the fireplace was sunk into a piece of stone.

'It wouldn't actually look *like* this,' Ferhan said. (They must have thought sometimes that they were dealing with imbeciles.) 'This is just to show you the quality of the finish.'

'Uh-huh,' I said, staring at the picture to try to work out what this meant.

'If you spent that sort of money it would mean you could have everything custom-made,' Joyce explained.

Charlie, slightly to my surprise, quickly did the sum and announced that he thought that we ought to go for the highest figure. This would mean a total construction budget of £470,000. He'd have to talk to Steve and, through him, to the Woolwich, but that was what he'd ask for.

So that meant that our income had plummeted at the same time that our expenditure was about to go out of control. Already we could smell in the air the money worries that would soon be settling like a fog around us. To go for the highest offered budget was incautious. But it also felt like absolutely the right thing to do, and, I thought happily (since I was not the one who somehow had to find the money), was absolutely typical of Charlie's unerring sense of what was important, and his commitment to doing the important things properly.

Joyce and Ferhan made sure we knew what we'd be letting ourselves in for. They reminded us that, aside from the construction money, we'd have to find fees for all the consultants we'd need to employ: a structural engineer, quantity surveyor, mechanical engineer, party-wall surveyors (ours and the neighbours'), services engineers, possibly computer and telephone consultants . . .

Besides that, their own bills would be coming in thick and fast. Their fees were set at 14 per cent of the project. By the end of June we would have paid 30 per cent of that, a month later 40 per cent. By September it would have gone up to 60 per cent and by the end of October 80 per cent. In practice this would mean that every six weeks or so, we'd be hit by bills of around £8,000.

We climbed down the steep and narrow staircase more soberly than usual that afternoon. Outside on the pavement Charlie stopped. He had to walk down to the Tube to go to another meeting, trying to pick up some work. He turned to me. 'We need to be aware,' he said, 'this is going to be really, really tough.'

6

About ten years ago, when I was younger and arguably crasser than I am now, I asked my friend Helen what her father used to do for a living (by the time we were talking, he was dead; I knew that much). Helen, as she was entitled to be, was disdainful of this question. I'm not sure whether she answered or not, because she has a very soft voice and mumbled her response and unless people speak in Estuary English or posh, I often have difficulty understanding them. (In my teens I went out for several years with a boy from Northern Ireland, a relationship that lasted so long mainly because I could rarely understand what he was saying.) But whether her mumblings were an answer or a suggestion that I should fuck off, she managed to communicate quite clearly that she thought the question was a) beside the point and b) simply not cool. She was right about this, in the sense that one is, or should be, friends with people for what they're like now, not because one likes or approves of their backgrounds. But Helen was particularly elusive; she was the most socially free-range person I'd met. I wanted to be friends with her, but had no more to go on than the evanescent scraps of the here and now. I wanted to know her in ways she didn't want to be known, ways she had transcended. I wanted to be able to fix her for myself. I wanted, I now realize, to know what her house had looked like.

Charlie and I had reached the detailing, Who Lives in a House Like This? phase of our design – which meant not merely making sure that we had enough rooms for our family, but that I had enough drawers for my T-shirts, meant looking beyond the arrangement of light and space to what sort of furniture the house would take and what it would say about the way we live and wanted to live.

We had studied the celebrity chandeliers in *Hello!* and the footballers' wives' boudoirs in *OK!* and considered ourselves reasonably expert in the semiotics of interiors. When the pictures of Cherie Blair at home with Carole Caplin came out in *Marie Claire*, we, like everyone else, were mainly transfixed by the bedcover and the lamps. And, like everyone else, we drew our conclusions. In this case, they were that people make the mistake of thinking that the Blairs are typically Islington – pretentious, effete, self-absorbed, lacking a sense of community and thinking differently from the rest of the country on almost any issue from immigration to the monarchy, bloodsports to homosexuality. But the reality – as demonstrated by that décor – is that the Blairs are not remotely Islington, but deeply, embeddedly Middle England.

Charlie and I had been announcing ourselves to the world through interior design ever since we stuck up our first posters with Blu Tack as adolescents. Now that we had our big opportunity, we felt we knew what we were about. And, of course, we were no more immune to Lawrence Llewellyn-Bowen-style fantasies of transformation-through-MDF than anyone else. All those house and garden makeover programmes trade on the well-understood principle that if your desk is a mess, you feel less in control than if you could only tidy it up. It seems to follow that if you were able entirely to reorder your living space, you might feel quite differently about everything, your whole life. The most monotonous and humdrum conversation about whether there are any clean tea-towels could easily turn into a coruscating cascade of wit or searching disquisition on the nature of identity.

So there was a lot at stake here. We had, besides, read enough sociology to know that as consumers (rather avid consumers, if the truth be told) we were engaged in a cultural project, the purpose of which was the completion of the self. 'I shop therefore I am,' as they used to say in *Marxism Today* with a revelatory air, although they were only articulating what girls from my bit of the East London/Essex borders had known for decades.

Our new living space would, we were aware, announce us as part of a community. When I was growing up, my parents were part of a G-Plan furniture, teak-shelving community. When I came back from Bahrain in the 1980s, many of the people I knew were in the grip of a Sloaney-chintz, Austrian blinds and yellow walls, anti-fitted carpets community. We could choose to become part of the sleek modernist set, as promoted by *Wallpaper* magazine, or join the band of Farrow and Ball fanciers, with old linen spilling evocatively out of whitewashed cupboards. We could in theory, anyway. In reality there was absolutely no question of anything Farrow and Ballish, of anything with the slightest taint of tweeness, snucking into an urban house designed by Joyce and Ferhan.

Modernists, it is sometimes noted, were the first architects to take control of every last detail of their buildings. God is in the details, Mies van der Rohe is supposed to have said, and probably did, although it's doubtful he thought it up, because half the people in history are also supposed to have said it, including Flaubert and St Teresa of Avila. But as Witold Rybczynski points out, all contemporary architects are Miesians in the sense that they share his preoccupation with perfection: whether modernists, postmodernists or deconstructivists, they expect to bring every aspect of the building under their control, replacing conventional details with designs that carry their personal stamp (in a Le Corbusier building, that might have meant windows that fitted directly into a groove in the concrete. In our house it means doors that go all the way up to the ceiling, benches to sit on, ceilings that are recessed all around the edges and seem to float – an effect technically known, I think, as a shadow gap).

Ferhan once insisted to me that what she and Joyce do isn't

about taste; it's about the organization of light and space, which she made sound like universal principles (and you can't quarrel with a universal principle). I wouldn't want to suggest that all their clients get the same thing; I know that in their time they have accommodated people wanting cupboards made of driftwood or wallspace for kitsch Hindu artworks. But although they claim to reflect their client's personality and lifestyle, I'm pretty sure that there are things that they would never do (only certain personalities need apply, perhaps).

As we sat down to do the detailing, Charlie and I were not about to choose how to announce ourselves to the world. Or rather, we had already made the choice. We were going to have what Joyce and Ferhan thought was right.

Probably this was just as well, because we didn't really know what we wanted, anyway. We knew what we didn't want, which was another Victorian house or the varieties of suburbia in which we'd grown up. Suburbia is one of those things that people feel entitled to be snooty about. One boyfriend from university who came to visit me couldn't believe how many Tube stops there were before you got to ours and another one kept marvelling (incredibly irritatingly) that the rows and rows of semis just went on and on.

What those boys didn't see was how different all of those semis were. My parents were always making improvements as we moved out along the Central Line towards Essex: tearing out 1930s wooden kitchens and installing streamlined units, knocking down chimney breasts and getting rid of fireplaces, building hi-fi units. Their friends were doing similar things with their properties. There was one couple with whom my mum and dad had a relationship apparently entirely predicated on DIY. Our family would no sooner hit on some new idea – woodchip wallpaper, say, or painting three walls of the woodchip in one colour and the fourth in another – than Shirley and Laurie would copy it.

With our faces all serious as we listened to our Judy Collins LPs, Elaine and I were convinced that all these DIY activities were not merely a very funny way of relating to other people but also a kind of displacement. We despised them as a way of shutting out life, as distracting sideshows to the real business of living, which was – well, something more like being a singer–songwriter. (It should be said that this coexisted with an equally smug sense that our house didn't smell musty, wasn't dark, didn't feel like it was inhabited by anything other than modern people, with state-of-the-art hi-fi.) Our parents were quite clearly suffering some kind of addiction, less harmful than alcohol or drugs – which were anyway unheard of in Wanstead and South Woodford in the 1970s – but all the same a way of blanking out life, because however much they improved our houses, the job was always incomplete: there was yet more improving to be done. It was only when I reached middle age that I recognized their need to feel always improvable, not settled, final. Of course they were improvable, and they knew it, and they didn't have that many ways of expressing it.

So I'm not snooty about suburbia: you don't need to live in a big house to have an expansive emotional life – although, in popular imagination, suburbia has (rather oddly, in my view) come to stand for anomie, for the sort of depression and anxiety that calls for frequent medication with valium or Prozac. You can find any number of almost reflexive uses of suburbia as a metaphor for spiritual death – *The Invasion of the Body Snatchers* in the 1950s, where alien body-snatching did time for the draining effects of living in the suburbs, through *The Stepford Wives* in the 1970s (in which scientifically contrived women are perfect by the conformist standards of suburbia only because they are mindless) to *American Beauty* in the 1990s, in which existential alienation surfaced like poison out of the tended, sprinkler-fed lawns. In Britain (unfortunately this came a bit late for me), suburbia is now recognized as sufficiently angst-inducing to be a perfectly respectable background for pop musicians, for Morissey and a whole crowd of subsequent furious, directionless, drugged-up youth. There is something about the popular idea of suburbia that is predicated on rigidity and a resistance to the questioning of social norms – yet it's difficult to understand why this should be so; difficult to identify, as the philosopher Carl Elliott has remarked, 'the precise characteristic of suburbia alleged to produce the kind of depression and anxiety that psychoactive medicine is needed to cure. Loneliness? Social conformity? A thoroughly modern kitchen? Whatever the problem of suburbia is, it is related to the problem of feeling homeless at the very place where you should feel most at home.'

Perhaps the easy linking of suburbia with spiritual death has its roots in nothing more complicated than that the majority of people live in suburbs, and that most of the questions that preoccupy most of us in the twentieth and twenty-first centuries are essentially domestic. We are not, as our ancestors might have been, primarily, or at any rate overtly, worried about our relationship with God or the state of our souls. As Carl Elliott notes: 'What we worry about (even when we worry about God) is ordinary life – the life of home and work and family. Should I get married? What should I do for a living? Should I have children? Where should I live? It is no wonder that the home has become a symbol of such anxiety: home is where it all happens.' Homelessness can stand in for meaninglessness, he adds, 'precisely because home, at least for late-modern Westerners, is where the meaning is'.

Not that this explained how we might avoid building a house that wouldn't make us think, 'God, is this it?' We just had to hope that Joyce and Ferhan had some sense of how to do it. When they had asked us early on whether we had thought we were brave enough for them, we had replied shakily (feeling not merely cowardly, but ignorant with it) that we hoped so. We knew we had to take direction. And they knew it too, because they'd seen the unthemed clutter in which we currently lived: they were probably worried that we'd suddenly decide we wanted rag-rolling and a sofa from World of Leather.

So they managed us. Their typical tactic was to show us, say, three samples of limestone and explain that one of them was the right price, the right colour and easy to keep clean. They did not mention that there were hundreds of other types of limestone, if we cared to look at them. They did not acknowledge that it was a bit daft showing us three types of limestone if two of them were not much good. What were we supposed to say? Oh, thanks, I'll have the overpriced one that doesn't go with anything else and is a bugger to wash?

At around the time we were embarking on the detailing phase of the project, they expressed their philosophy in an interview in the *Financial Times*. Ferhan quoted Frank Lloyd Wright, 'You should give your client not what they want but what they need.' Once they'd convinced each other, they told the journalist, it's easy to cajole 98 per cent of their clients.

The first time they mentioned concrete, I made a face. We were discussing the garage, specifically how it would appear across the garden from the house, and Joyce said, 'Well, you could, for example, look across at a beautiful piece of concrete.'

She caught my expression. 'What? You don't like concrete?'

I did not like concrete, no. It made me think of multi-storey car parks, and blocks of flats thrown up in the 1960s, with stained walkways and pissed-in corners, streaked facades and crime-scarred corridors. A beautiful piece of concrete seemed to me tautological nonsense.

(Not that I subscribed either to the idea that slum clearance and rebuilding was a patronizing spree by architects who were heedless of social memory in the places they were building, because they themselves lived in Georgian terraces somewhere else. I knew it was never that simple: I remembered my grandparents being slum-cleared from an industrial cottage in Hackney Wick to a new council flat in Stepney Green and being impressed, as a small child, by their excitement at its cleanness and functionality, by their delight at not having to go outside to the toilet or on the bus for a bath. All the same, this was a new century. I was not my grandparents. I could see the value of Victorian houses, especially with the addition of a Poggenpohl kitchen. And as far as I was concerned, the whole point of the welfare state had been to save me from living in a concrete building, rather as it had saved me from following my aunts into a garment factory.)

Charlie, meanwhile, had grown up in a town built wholly of concrete – Basingstoke – to which his parents had moved when he was seven, when the place was still largely a utopian vision – overspill containment in delightful countryside – in the heads of government planners. He had picked his way through building sites en route to school, had trailed, in disconsolate adolescent fashion, past modernist office blocks, been beaten up in shiny new underpasses. He thought of Basingstoke as soulless, a half-baked attempt to impose community through concrete, and it had largely inoculated him against the material's charms.

Joyce dropped the subject and I assumed I had despatched the eccentric concrete idea. And then, at the first detailing meeting after planning permission, she and Ferhan announced that they were thinking of what they called 'an envelope of fair-faced concrete' at the narrow ends of the house, the glass walls strung between them.

Fair-faced concrete sounded like another contradiction in terms. But, as ever, when there was a tricky decision to make (or for us to see the wisdom of) we didn't have to worry about it right away. They just wanted to let us know that they were meeting one of the country's leading concrete experts to discuss the possibility of such a thing: and look, let's be realistic, it might never happen; it was probably unfeasible. They showed us a book that the concrete man (his name was David Bennett) had written. It went into great detail about light grey cement plus fly ash filler, Portland cement and pulverized ash fuel with a water-cement ratio of just under 0.5, ground-granulated blast-furnace slag cement with yellow sand and river-dredged coarse aggregates. It was eloquent on the subjects of release agents, poker vibrations and aggregate–cement ratios.

But, like they said, in-situ concrete (brought to site liquid, rather than in pre-cast chunks) is very rarely used in this country for buildings of the size of our house – as opposed to National Theatre size – because it easily goes wrong. And it's expensive, though they didn't say this. And people don't much like it. It reminds them of multi-storey car parks. Anyway, there was no point in our worrying about it now because their meeting would probably lead to the conclusion that it was impractical.

Joyce and Ferhan announced in May, before we got properly stuck into the detailing, that they wanted to get a fix on our philosophical position as regards environmental sustainability. We thought for a minute and replied that our position was that sustainability should be a constant concern but not our primary objective. We were all for environmental soundness, but we didn't want to build a yurt.

This seemed to be the right answer. There were, they said, various sustainable elements they might be able to incorporate into the building, such as solar heating, natural ventilation, the collection of rainwater and recycling of bathwater. They were most optimistic about the last: at the very least, they said, it ought to be possible to use the bathwater on the garden. They would look into the various options and report back.

We never heard a thing about environmental sustainability again. When I asked some months later what had happened to the bathwater scheme, Joyce said airily, 'Oh, it would've been way too expensive.' And perhaps it would, although presumably any proper cost-benefit analysis should have involved us, since there were long-term savings to set against capital outlay. Perhaps, I thought darkly later, the environmental add-ons might have entailed physical add-ons – bits of building that would stick up, or out, or even be *curved*.

Anyway, our relationship with sustainability was not unlike our relationship with gyms: goodwill lasting approximately a fortnight, followed by lasting guilt. My mental dealings with Kevin suffered somewhat here (in my head, he was still popping round for a biscuit and to admire our progress). He is rather keen on environmentally sound buildings; I felt his disapproval and didn't invite him for a bit.

Much later, I gained further insight into the non-existent sustainability, when I attended a talk Ferhan gave at the Royal Institute of British Architects (RIBA). Someone in the audience asked whether, if you were undertaking a large-scale project, it didn't make sense automatically to incorporate sustainable elements? He hadn't, he said with some concern, heard much about this in the initial presentation. Ferhan responded by talking about the stringency of building-control regulations, the need sometimes to get round them, and Azman Owens's skill at getting building control enforcers to see their point of view.

Clearly, if we'd wanted an environmentally radical building, we should have hired different architects.

By May, all the children's bedrooms had moved upstairs, following a sort of game of musical spaces that also gave me a proper study in what had until then been the den (which was now in Hen's former room inside the front door, next to the kitchen). This shift also had several less beneficial consequences, not all of which were immediately apparent. The den shrank. So did the store room. For months afterwards, whenever I couldn't think of a place for something (vases/shoe polish/brooms) I'd say blithely, 'Oh, well, we'll keep it in the store room,' not realizing that the storage space in the store room had now been reduced to roughly the size of one kitchen cupboard. Joyce and Ferhan never, in fact, referred to it as the store room again, but always as the utility room. And really, they should have called it the narrow utility corridor.

Meanwhile, it was time for us to make some effort. Joyce and Ferhan set various bits of homework: measuring the height and the length in shelf space of our books; compiling a list of what we needed to keep in the kitchen cupboards; taking home a rough wardrobe plan and testing it against the reality of our clothes. Among other things, this wardrobe plan contained a shelf marked 'seasonal handbags'.

'What on earth are they?' Charlie asked.

'They're handbags for different seasons,' I said superciliously, showing off in front of Joyce and Ferhan, though I had never owned such a thing.

'Very important,' added Joyce.

Until then, my tactic with handbags had been to buy one and use it relentlessly until it developed a hole or the lining was so torn that my keys regularly got lost in it, when I would stuff it down behind the chair in our bedroom, just in case I wanted it again in future. Needless to say, I had never yet retrieved a bag from this graveyard of leatherware. But – and this shows just how useful architects are – as soon as I had the notional shelf for seasonal handbags (not even the shelf itself) I found I needed the bags to go on it. I now have a number of seasonal handbags, taking up even more than their allotted space in my wardrobe.

Meanwhile, to help with my thinking about the kitchen, I went to visit Hugo, who, since he once worked in a professional kitchen, not only has a proper *batterie de cuisine* like they tell you in cookery books, but knows how to arrange it. He let me poke about in his cupboards without once having to say, 'Well, of course the balloon whisk isn't really meant to live behind the sugar.'

Style cramps life, and life erodes style, the biologist, inventor and 'long view' advocate Stewart Brand has written. And I'd be inclined to agree, except that it isn't true of Hugo, who not only managed without having on show all the things that cluttered up my surfaces – bread boards, food mixer, microwave, toaster – but didn't actually have much inside his cupboards either. About once a month, he explained to me seriously, while I admired his arrangement of wooden spoons, he goes through the kitchen and moves out anything he isn't using regularly. I had cake tins in my kitchen cupboard that I'd used once, thirteen years ago; I had a jelly mould that was rusty but which I kept simply because I didn't have another one. I went straight home and filled eight bin bags with kitchen rubbish.

Other useful tips I picked up from Hugo's kitchen and added to my Kitchen List for Joyce included:

china and glass near dishwasher
shallow drawer next to hob for cooking implements
deep drawers underneath it for saucepans
olive oils, salt, pepper, nutmeg, Marigold bouillon, etc. at convenient height by the hob
a shelf in a cupboard for anything that people put down on your surfaces – bank statements, letters from school, toys from party bags – which, if not removed by owners within a week, will just be thrown away(!)
pull-out bin in a drawer underneath the main chopping surface, preferably stainless steel for easy cleaning

For some reason, probably to do with the number of times the kitchen cupboards were reconfigured, a door, rather than a drawer, was finally allocated for the bin, which rather undermined the idea of carelessly sweeping your choppings towards you (i.e. now on to your feet). It also limited the size of bin to one that could hang on the inside of a door – in other words, to one wholly unsuitable for a family that produces a small landfill of rubbish every week. In the end, I tore off the bin in disgust so I could use the cupboard for something else, and today my bin is a black plastic bag on the floor. Charlie hates this and is always muttering about getting a 'proper' bin but I like the plastic bag arrangement. At least you can chuck it outside when people are coming.

Our kitchen budget was £17,000, including appliances, which is not excessively high in these days of extendable mortgages (and would buy you, in fact, about one-fifth of a fancy German kitchen). This meant that all the cupboard doors could be custom made, but what went inside them had to come out of a catalogue of kitchen fittings. (Hence, perhaps, the difficulty with the drawer-bin.) One of Hugo's clever kitchen things was a knife drawer with a wavy base, for keeping knives separate from one another. Joyce and Ferhan ignored my plaintive references to this until it was clear I wasn't going to stop going on about it, when they said I should look into getting it made myself. But I knew no one who carved wood into wave devices. I realized, what's more, that one of the great pleasures of employing an architect is not having to find such a person. Not having to do very much work at all, in fact, beyond opening one's mind to the possibility of seasonal handbags. Not having to trail around shops, bewildered by the hubbub and the choice and the necessity of making a decision.

I had hated trailing around shops in which you had to make decisions, rather than buy, ever since I was a child and my parents had displayed an incomprehensible ability to spend whole Saturdays in the Bakers Arms Carpet Centre, while I gnawed my knuckles to stop myself from screaming and feared that I might simply, before I had been allowed to grow up, go mad. So it was restful to decide that you were going to trust someone else when they said you should have either this door handle or that one. (Joyce and Ferhan showed us two stainless-steel handles, helpfully attached to pieces of wood, and we had to try them and say which one was more comfortable, which would then be used around the house. The handles looked remarkably similar, but it's possible that even here we were directed; my recollection is that one was significantly easier to use than the other.)

Until then, I had always thought people who employed interior designers were a bit odd: why would you want to be bossed about by someone who thought their taste was superior to yours? But, somehow, employing an architect was different. For one thing, we had little sense of how our as yet non-existent house would look, or the way it would work. For another, we were clueless. It was getting on for ten years since we'd done up our house in London Fields and we hadn't done it properly even then: we were still living with a decorated fingerplate on our bedroom door, white china with little roses, left there by the previous owner, who had originally shown us the bedroom saying, 'It's a very sexy room, don't you think?' so was evidently not entirely sane. And we were dealing with universal principles, which Joyce and Ferhan had access to and we didn't.

But I did insist on an American fridge. I had only stayed in one American household where such a fridge was in use and it was mainly used to store something called iced tea, which came out of a bottle and tasted nothing like tea, plus a large assortment of fizzy drinks and snacks. (The stuff my children call 'food', as in 'There's no food in the house' when the house is actually full of lettuce and apples and cheese and bread and things that do admittedly require some cooking.) The teenagers in the American household would gravitate to the fridge whenever they entered the kitchen, so you mostly ended up having conversations with their bottoms as they fished for snacks. This particular fridge, then, was not a great advertisement for anything, and Charlie even wrote a column for the *New Statesman* citing the big fridge as the root of the American obesity problem, because it had to be filled with food, which then had to be eaten. Ban big fridges and you could slash healthcare costs, was his policy.

But since then my sister had acquired an American fridge, which was impressive mainly for incorporating an ice dispenser that sent big chunks or delicate slivers of ice, depending on your mood, crashing into your glass. I am shallow, but I wanted one of these. Joyce and Ferhan drew a normal-sized fridge on their original plans because the proportions were more elegant. But I was adamant. The kitchen had to be drawn and redrawn several times, but, finally, after several detours – an extra sink, a 'coffee station', whatever that is, a counter top at right angles that quickly got rubbed out again – our enormous ice dispenser found a place.

Architects may have started out as top tektons, but at some point they became – certainly in popular imagination – less like builders, more like artists. The trouble with art, though, is that it is essentially non-functional, it actively aspires to be impractical, and it despises the conventional and manically pursues the new. None of this seems conducive to the creation of buildings that are easy to live in. So when Joyce and Ferhan simply took for granted that we would have taps made by a company called Vola, 'Because we like them and we use them a lot', we were torn between trusting them – which we wanted to do because we'd already invested a lot in this emotionally – and being very alarmed. The Vola taps looked OK, from their picture, in a way that was scarcely there, which was presumably the point of them; but fundamentally, we could only hope that this house would be habitable, and not only by architects with their funny ideas about two toys at a time and not producing much rubbish, but by us. It is the job of architects to save clients from themselves, it has been claimed; in the constant tension that this entailed, we had to hope we weren't being saved from ourselves in the manner of one of those religious suicide cults.

While we – OK, they – were designing the house, I read one

of the few books that has ever been written about architecture from a client's point of view. Suzanne Frank is an architectural historian and the owner of Peter Eisenman's House VI, which was completed in New England in the summer of 1975, and her book, published in 1994, *Peter Eisenman's House VI, The Client's Response*, or at least my copy of it, came with what may well be the best errata slip ever:

Cover: The drawing is printed upside down. It is printed correctly on page 105.
P.105: The drawing in the upper right-hand corner is printed backwards.

As soon as you start looking at the drawings, though, you realize why the publishers had such a struggle. Eisenman is a deconstructivist, I suppose: at any rate, his architecture is abstract, literary and self-referential. Not all of his previous V houses had been built, which makes sense, because they approach the condition of conceptual art. (Eisenman once told *Newsweek* that his buildings were 'designed to shake people out of their needs'.) And, like conceptual art, the conscious revelation is only available once; there has to be something more than a cerebral point – has to be some mystical or emotional experience – for the work to remain interesting. And the main emotion that it seems to me Eisenman's House VI would generate is irritation.

I hoped that Joyce and Ferhan weren't trying too hard to shake us out of our needs, because the consequences for Suzanne Frank had not been great. (This would not necessarily be her opinion.) Her worktops are the wrong height, to fit some complex system of planes that Eisenman devised, meaning people had continually to bend over. Conversation at mealtimes is virtually impossible because of a column that descends into the only space available for a dining table (for no structural reason). There is only one bathroom and you have to go through the master bedroom to get to it. Eisenman's system of floorlighting required that Frank and her husband sleep in single beds.

(Eventually, greatly daring, they refused to put up with this any longer.)

In an Afterword, Eisenman explains:

In House VI, the experience of the physical environment does not lead to any mental structure – the experience is, in fact, quite the reverse. Once the conceptual structure is understood, it detaches itself from the initial physical experience . . . In House VI, a particular juxtaposition of solids and voids produces a situation that is only resolved by the mind's finding the need to change the position of the elements.

It is not entirely clear what he means here, but I think it's that he put things in the 'wrong' place to make us think about whether the 'right' place is merely culturally conditioned. Too bad, of course, if it isn't. Too bad if the right place has been arrived at by experience, because the wrong one is bloody uncomfortable.

Joyce and Ferhan were of course strictly form-follows-function architects: there would be no staircases going nowhere to make us ponder the *idea* of a staircase in our house, no doors that didn't close the opening to the master bedroom, simply to be the sign of a door, to make us *think* about doors. Even so, we had stopped worrying about how we'd feel if the house was all about us and people didn't like it, and moved on to what it would be like if the house was all about Joyce and Ferhan and we didn't like it.

How much did they know about us anyway? Ferhan once told me that she'd always thought of me as a very sexy woman, which marked her out as a person of great insight and discernment, although I suspected she just thought this because I'd told her I had an underwear weakness and needed plenty of space for my lingerie.

Just as we could never have turned the confusion of our dreams and anxieties into three-dimensional space (even if Charlie did buy an architect's pencil, identical to Ferhan's, and brought it to meetings), they couldn't know us from the inside. Dwelling, says Ivan Illich in *The Mirror of the Past*, 'is an activity that lies beyond the reach of the architect not only because it is a popular art; not only because it goes on and on in waves that escape his control; not only because it is of a tender complexity outside of the horizon of mere biologists and systems analysts; but above all because no two communities dwell alike'.

I loved the detailing stage of the design, looking forward to meetings at which I was invariably presented with something that opened up possibilities of a new way of living – a bench along the kitchen wall, where the family would sit and talk to me while I was cooking; floor lights that would glow softly when I came in on dusky winter evenings; a trough in our bathroom in which we would grow tall bamboo. But at the same time, the experience I was most reminded of was being on a camel in the Sinai which bolted down an enormous sand dune.

While we were waiting for the planners to turn their attention to us and floundering around with Tom Tasou, Elaine had quietly bought the house next door to hers in London Fields and employed an architect to knock the two into one – lateral conversion, it's called (offering magazine sub-editors a rarely missed opportunity to use the headline 'lateral thinking' over articles about it), so by now there was a lot of showing of drawings and comparing of progress. I could tell she thought I was a wimp for simply looking at a picture of Vola taps and saying, 'Yeah, fine, OK,' without demanding to see a whole bunch of tap catalogues. I worried sometimes that I was like Peer Gynt peeling the onion in the play, and that if Joyce and Ferhan only peeled away for long enough with their Vola taps and Guido beige limestone I would be exposed as nothing, a universal principles victim.

Henrietta and Freddie had a much clearer sense of themselves.

Fred came to an early detailing meeting at which kitchen materials were discussed: stainless-steel worktops and cupboard doors, timber, glass. Even if you knew nothing about architecture – and Freddie loved buildings – there could be no doubt that the house was going to look simple, cool, sleek. As we were coming down the stairs from the office, he turned to me and said: 'Can my bedroom be gothic?'

Hen later produced a picture of what appeared to be a penthouse flat, which vaguely reminded me of the place where Michael Douglas had lived in the movie *Wall Street*, with a mezzanine approached by a spiral staircase, on and under which at least twenty people were partying. Would it be OK, she asked, if we got Joyce and Ferhan to make her bedroom like that?

Harry and Ned were too young to have views. Harry regarded the project as an eccentric parental preoccupation roughly on a par with reading newspapers at breakfast or checking emails i.e. tiresome but not actively harmful. Both of them were still young enough to trust that not only were our intentions benign, but the results would be too. The only aspect of the project in which Harry displayed any interest at all was the tree-house. 'It can be really simple,' he announced. 'I don't mind what it's like, as long as it has a television.'

We did insist on some things. Initially there were no desks in Henrietta and Freddie's rooms. In their enthusiasm to think about our long-term occupation of the house, I think Joyce and Ferhan had already despatched the two older children to lofts in Hoxton, or at least to flats above the betting shop in Kingsland Road, even though Hen was in the thick of A-levels and planning four years at university and Freddie hadn't even started GCSEs.

I was also firm about lighting. I felt I knew something about this, on account of there being a chi-chi lighting shop in Islington. Admittedly, I had been asked by the staff of this shop that no member of my family ever again attempt to dismantle one of their pieces, after I'd returned a standard lamp in seven sections.

But I didn't care for the pendant lamps Joyce and Ferhan were suggesting to hang above the kitchen table and in the corner of the sitting room and volunteered to find my own. The minutes of that meeting merely recorded that the light fittings had yet to be decided.

By the end of the summer, we'd more or less finished the detailing. The children's bathroom would have a red rubber floor and stainless-steel fittings; their bath would be a tiled tub with right-angled sides, like a rugby bath. Our bathroom would have limestone floors, a timber-surrounded sink, the bamboo trough and the skylight that opened. For the timber, which was now everywhere – shutters, cupboards, doors, desks – Joyce and Ferhan were keen on something called iroko, which we had never heard of, but which, they assured us, weathered to grey, or, if treated, to a rich reddish-brown. They didn't actually say so, but a crucial factor in their enthusiasm was that it was cheaper than pine. I asked if it was sustainable, and they said yes; I questioned whether you could be sure where it came from; they said you could. And I wanted to believe them, because they wanted to use the iroko, and because my trust had become almost a self-fulfilling prophecy: I trusted, therefore they must be trustworthy.

We had not, like their best and favourite clients, gone off on a three-month photographic assignment while they polished off the building. It may be true that you can't make art by committee, but architecture is a uniquely social art, a process of acknowledging and submitting to context. Architecture must respect the surrounding buildings, the area, the lives of its future inhabitants. And if it's true, generally speaking, that artists suffer from competing needs to get enough outside stimulation and to be left alone to develop their own ideas, then architects probably suffer more than most, because most of their external input is coming from people who have very muddled attitudes to them. As a client who wants a fabulous new building, you have quite a lot invested in believing that your architect is an artist, a creative genius possessing knowledge, intuition and inspiration. At the same time, you are the person with the most to lose should your architect get carried away with the notion that architecture is ultimately about art, rather than about life.

7

There are two ways of appointing builders, Joyce and Ferhan announced in May. You can put the project out to tender and see what happens, or you can pick a builder and negotiate the price. For reasons that I never fully grasped, and that seemed, when I thought about them, counter-intuitive, on this project they favoured the second.

The builders they'd chosen were people they'd worked with before and trusted. Most importantly, they explained, these builders had an absolutely brilliant site manager, which would apparently make all the difference. Unfortunately, being so good, they were very busy, and we'd have to wait for them until late September. But if the securing of a brilliant site manager was important enough to dispense with giving these particular contractors the bracing uncertainty of getting the job as they priced it, and not to check that someone else out there couldn't do it for half the price, who were we to argue?

Of course, Joyce and Ferhan's resolution to use a particular contractor would have made a whole lot more sense if there was something else going on. If they had an ulterior motive. If, say, these builders had some particular expertise that we needed, such as being especially good carpenters, and good carpentry was crucial to, for instance, concrete.

The meeting with the concrete man had gone well. Joyce said he was 'interesting' and she and Ferhan took to dropping concrete into the conversation more and more: 'in-situ concrete', they would say, perhaps hoping this would confuse me. (As far as I could work out, in-situ concrete was brought to the site wet and gloopy, poured into a wooden mould and left to cure. When it was dry, the wood was peeled off, leaving the set concrete imprinted with whatever had been on the wood – a lot of splintery grain, in the case of the National Theatre; not very much at all, I hoped, in the case of our lane.) They showed us a picture of a building in the new Thames Barrier Park, which they said was beautiful. We peered at it. It looked like a bus shelter.

I told my friend Richard Stepney, who cuts my hair, that our architects were trying to persuade us to have concrete. He said, 'How would you react if I suggested a haircut I'd never done, which had been out of fashion for decades, that I wasn't sure I could make work and most people thought looked ugly?'

He had a point. It all seemed absurdly risky. In the discussions with Joyce and Ferhan, I would always say, '*If* we have concrete.' There was a meeting early in July at which I did this so many times, with so many caveats – it had to be smooth, it had to be pale, it had not to have ugly wet streaks down it – that Joyce was obliged to accept that the usual methods weren't working.

She was forced to send us a letter.

> Dear Charlie and Geraldine [she wrote],
> Following our meeting on 5 July 2001, please confirm that you would like us to progress the design of the house using in-situ concrete walls for the load-bearing masonry walls.
>
> In the meeting we discussed the finish of the concrete. We explained that you should expect the concrete finish will not be free of imperfections of colour or texture. Each project will have its own variations depending on the weather, the contractor, the additives, the concrete mix, etc. However, the contractor has expressed confidence in building the concrete walls and we will ask for test panels to be constructed so that we can review the finish that will be achieved.
>
> We will advise the ready-mix company that you prefer a

lighter-colour concrete and we can review the samples with various colour additives. As discussed, to minimize the grain effect on the concrete we will specify birch-faced plywood instead of shuttering plywood.

We await your instructions,

Yours sincerely,

Joyce Owens.

A few days later I sent back a reply, unhelpfully larded with neurotic equivocations:

Charlie and I are happy to progress the design of the house on the basis that we will go for in-situ concrete walls. I have a couple of reservations/anxieties however. The first is over colour and blemishing. I sometimes find concrete oppressive and lugubrious. For this reason I am keen that the finish should be light (in the sense of pale).

I accept that unevenness and blemishing can be charming, as it is in wood. What I dislike is the water-staining effect, which makes buildings look as though they have a damp problem. I don't know whether this occurs at the time of construction or subsequently.

And then I added some further worries about insulation and getting the concrete poured evenly, which were just reiterations of things Joyce and Ferhan had said could go wrong. What I hoped to imply here was, 'There seem to be an enormous number of pitfalls. I am not qualified to judge, these being technical matters, but I expect that after you and the contractor and the structural engineer have given them proper thought, you'll decide it's just too risky.'

And then I sat back and nursed my misgivings and waited for them to find out it couldn't be done.

★

In August, we went on holiday to Colonsay, an island in the Inner Hebrides – or rather, five of us did. Hen was inter-railing around Europe with Matthew. Already the family that Joyce and Ferhan were supposed to house in all its heterodox untidiness was pulling apart, like planets in space.

Hugo and Sue spent a fortnight on Colonsay every year. We were becoming like my parents and their friends Shirley and Laurie. But whereas Hugo and Sue always stay with the laird, at a house party in a magnificent manor house with vast kitchens (and a professional cook) and lovely gardens stuffed with interesting trees brought back from foreign parts by laird ancestors, we stayed in a damp cottage with lumpy beds. Colonsay is wild and remote and fringed by heart-stoppingly beautiful beaches of pale sand, rocks and clear, if icy, water. But one room in our rented cottage was too damp to use: you had to keep the door shut to prevent the musty, wet smell from escaping and wafting through the rest of the house, laying its insidious creeping fingers on the broken-spring beds.

It wasn't particularly easy for any of us to get a good night's sleep, nor advisable, since a whole night in the same position meant persistent neck or back ache the following day. Charlie lay awake night after night on the miserable poking mattress worrying about money. The tax bill was due in January – which, now that we were at the end of the summer, didn't seem so far away. In July 2001, we'd paid 40 per cent of Joyce and Ferhan's fees; by October, we would have to pay 80 per cent. In practice, this meant bills of £10,000, £7,000, £3,500, £7,800, £2,500, £7,800 hitting us at alarmingly close intervals through the school holidays and the start of term. The structural engineer, Brian Eckersley, had also sent in his first bill, for a couple of thousand. I could see the point of Brian, who would make sure the building didn't fall down, but when we'd agreed to appoint him, I'd said nervously, 'There won't be any more hidden costs, will there?' I'm not sure what Joyce and Ferhan had answered, but, since then, we'd acquired a party-wall surveyor to make sure we didn't damage the walls on either side of us, plus party-wall surveyors to act for each of the neighbours (though we had to pay for them); a services engineer and the structural engineer's drawing expert. Already it felt like we were employing half the building and allied tradespeople of Southern England.

We'd had to plunder our savings to pay Tom Tasou and now, suddenly – whoops! – it appeared we had no money.

Charlie spent the long, uncomfortable nights on Colonsay fighting the cold he was getting from the damp and making mental lists of people he could approach for work. When we got home he transferred the lists on to paper, and made the calls. He persuaded the technology-in-local-authorities organization to pay, but it was obvious the relationship was coming to an end. Then came the terrorist attacks of September 11th and, in their wake, many fears, one of them of global recession. It was not a good time to be looking for work, or, for that matter, writing a book about optimism. By the first week of October, it was impossible to get money out of a cashpoint; I paid for food with a credit card that gets paid off automatically the following month, and hoped that somehow there would be something in to cover it by then.

The firm of contractors Joyce and Ferhan had chosen was called Varbud. It was owned by Ramesh Patel, who'd started it in 1980. Patel wasn't really Ramesh's name; his father had changed it from Visani when he'd emigrated from India to Kenya. Lux, Ramesh's wife, who was a Patel before she married as well as afterwards, told me that her father claimed that he'd changed *their* name from Halai because it was easier for the British to pronounce. But I'm not sure how much easier Patel really is than Halai, or Visani, for that matter. Perhaps at some stage there was also a desire to escape other things. The name Patel, Lux told me dismissively, 'is meaningless' – a caste name and the most widely used surname in the world. Say 'Visani' or 'Halai' to a Gujarati, on the other hand, and they would immediately presume to know everything important about you: what group of villages your family originally came from, who your relations were, whether any of them had ever behaved scandalously.

Ramesh was born in Kenya, the youngest of three sons. When he was nine, his father decided to go home to India and fulfil his ambition of building India's first drive-in cinema (which he did; it's still the largest drive-in cinema in the country). After his parents left, Ramesh stayed in Kenya with his older brothers, because the education was better (Ruda, the eldest, who really brought him up, now does all the ordering for Varbud).

Ramesh moved to Britain when he was eighteen, to study quantity surveying at Middlesex Polytechnic. Not long after he'd graduated, he went camping in Europe with his friend Yogi and, sitting in their tent, they decided to start a business. Neither of them really had any money, so they each put in £50. There was never any question but that the business should be a construction company. The Patels were a farming caste originally but, as the land became less labour intensive, they diversified, and what they mainly diversified into was construction. According to my Gujarati friend Dilip (himself a sexual health counsellor turned garden designer), Gujaratis are particularly renowned for their joinery.

Initially, Ramesh and Yogi ran Varbud in their spare time, in the evenings and at weekends, organizing labourers they knew to knock out fireplaces or build walls, and carpenters to make shelves. In 1982, an acquaintance from the Gujarati community asked them to build a house in Kingsbury. They gave up their day jobs. The project was successful, word got around, and from then on, they were never without work.

By 1983, when Ramesh called Lux Patel's father and asked permission to go out with his daughter, he was a pretty good prospect. Lakshmi (pronounced Luxshmi, but people were always getting it wrong) was twenty, and a nurse. Pretty and smart, she had already made it very clear she had no intention of having an arranged marriage like her sister. The older Mr Patel advised the younger one to ask her out himself. Initially, she was disdainful. 'I've never seen you,' she objected. 'If I don't like you, I shall probably vomit.'

When they married in 1984, they bought a house in Kenton, in West London. Varbud set up in the box-room, and when Lux had two sons, she began doing the books and chasing payments.

By 1990, they'd outgrown the back bedroom. Varbud moved to a warehouse on an industrial estate in Perivale, with room for a joinery shop and store room downstairs and offices upstairs. Today, the business employs forty people full-time and rarely takes on contracts worth less than £150,000.

In a sector that relies heavily on trust (you can't afford to employ plumbing subcontractors who don't do the job properly) the Patels have the advantage of coming from a community where reputation is prized and carefully monitored. It matters what kind of a family you come from, what its standing in the community is, how much it gives back, whether its members behave themselves and live good, useful lives. In this community, Ramesh is highly respected. He employs a lot of people, the work they do is of high quality, he is devoted to his family and he spends a great deal of his time on charity work. Since his middle brother died, he also oversees the cinema business in India (his father is in his mid-eighties and has had several heart attacks), which necessitates at least three longish trips a year. When Lux told me that her kitchen at home wasn't finished after five years, I thought, 'Hmm, how interesting! Obviously doesn't bother her much. It's not embarrassing, even though she and Ramesh run a building firm . . . There must be other values here, more important than domestic display, to do with family and community and respectability.' I did not draw the more obvious conclusion, that Varbud just takes a long time.

Still, David Bennett, who had insisted on going to see their workshop, had been highly impressed by their carpentry. He'd decided they could be trusted to make the shuttering, with the proviso that everything must be cut and prepared in Perivale – 'We want no cutting and hacking on site.' Each piece of birch ply would be used three times, but in between it must go back to the workshop to be cleaned down and have release agent reapplied. This attention to detail was crucial. Tadao Ando, the Japanese architect whose simple, light-washed buildings are in large part responsible for the recent reassessment of concrete, has claimed that 'the high standard of Japanese concrete building is based on our woodworking heritage. Concrete moulds must be made with great care and precision to produce a clean and perfect surface. As a matter of fact, I used to employ joiners to make my moulds.' Well, we would have joiners with heritage too.

Joyce and Ferhan showed us three types of contract, relating to different sizes of job. Although our project was borderline small to medium, they thought the medium-sized contract would give us better protection. They'd have to make some adjustments to the standard format, which would then have to be approved by our lawyer and Varbud's lawyers; once they'd all finished arguing, we'd be ready to sign.

'But we're due to go on site next month,' I yelped. 'That'll take ages!'

'Oh, you don't have to sign before you start,' Joyce said airily.

'But what if something goes wrong?'

'The contract's still valid. It's quite common not to sign the contract until the job's finished.'

This seemed to me a pretty amazing contract, that was binding even though you'd only ever glimpsed it upside down on someone's desk. I tried to work out what would be the point of signing it after the job was finished. And presumably there wasn't one, because we never heard of it again.

The important thing seemed to be the letter of intent, which Joyce and Ferhan wrote on our behalf, and faxed over for us to sign and send to Varbud. This stated the contract sum (£474,000) with a defects liability period of twelve months. (This meant we would withhold 2.5 per cent of the money every time we got a bill. After a year, the builders would return and make good the things which had gone wrong – and Joyce assured us things always did go wrong: doors would drop and cupboards would shift – and we'd pay them what we owed.) The letter also stipulated the start date. Initially, we'd hung on for Varbud until September. Imperceptibly, this had turned into October. Now it had crept into November, with the 12th frequently mentioned as a good day. Varbud were still claiming that they'd start on the 12th, but the first week was now to be 'preparation in the workshop', whatever that meant (how hard do you have to prepare to dig up some ground?) and they would move on to site on Monday the 19th.

The job would take twenty-eight weeks, plus two weeks off for Christmas and a week for Easter – which, from 19 November, took us to 30 June 2002. After this we were entitled to something called liquidated damages: compensation of £550 per week, which was roughly what we estimated we'd have to pay to rent somewhere.

Meanwhile, Varbud sent us their costings. We sent them back, querying things, and demanded that they break them down room by room, so we didn't just have a large sum (a very large sum, in fact) for 'joinery'. But even after they'd revised their estimates and we'd questioned almost everything, even after the resulting adjustments had been made, we couldn't get the cost down below £480,650. And Varbud still hadn't priced for the bin store that we had to build as part of our agreement with Tom Tasou (which would take the price up to £483,400). There were a number of items on the list that were still provisional; Elaine, whose lateral conversion was now zinging along, said that a good rule with things marked provisional was to double them. One of our provisionals was, worryingly, the kitchen. That meant we were already more than £8,000 over budget, with the likelihood of some things coming in higher than estimated and no contingency. Joyce said a contingency of £20,000 would be safe.

In September, Charlie and I met Joyce and Ferhan at their offices to try to resolve the problem of not having enough money.

'We could solve this at a stroke,' Joyce said, 'if we got rid of the poured concrete.'

I stared at her. Ferhan had opened a book and was pointing at pictures of pre-cast concrete blocks. Joyce was talking about render, and about how she'd really wanted to put zinc panels on the house she'd built with Bill but she hadn't been able to afford it. I swallowed hard and blinked. My mouth had gone slack. My eyes were filling with tears. I was going to cry over Ramesh's costings, to drip mascara over the white coffee cups and the interesting black tile that Joyce and Ferhan used as a tray.

'Excuse me,' I muttered thickly, banging down my coffee cup and scraping back my chair.

In Joyce and Ferhan's bathroom I blew my nose and stared at myself in the mirror. My eyes were puffy and watery and my face was hectic with red patches. I was like a child in the intensity of my wanting, in the outrage and injustice I felt in not being able to have.

I couldn't understand it. Why was I so emotional? I didn't even like concrete. I'd been against it all the way through. I was the one who was supposed to be ambivalent; I'd been dragged along because everybody else was so insistent. Concrete was risky, difficult, contentious, made people laugh incredulously when I mentioned it. It was against my better judgement. And, it now seemed, it was the essence of the building.

The truth was that I'd been educating myself, on the sly, making a concerted effort to talk myself into it. In particular, I had applied myself to the work of Tadao Ando, or rather, to pictures of it, because the work itself was all in Japan. And the pictures were certainly very beautiful, showing shafts of light and slivers of shadow on simple walls. Some of the photographs were taken in the mauve light of a summer evening; others inside in the middle of the day with bright sunlight branding stripes on the concrete. In the pictures, what you actually saw was light and shade (things that Tadao Ando has claimed he considers architectural materials). 'The concrete that I use does not give the impression of solidity and weight,' he said. 'My concrete forms a surface, which is homogeneous and light; the surface of the wall becomes abstract; it is transformed into nothing and approaches infinity. The existence of the wall as a substance disappears.'

Not only did the concrete not look like concrete, it attained an abstract state of non-being, even approaching the condition of infinity. It was hardly there at all. How could concrete like this be threatening, when it was practically absent? It was too busy being metaphysical.

I'd also discovered that concrete had a history pre-dating tower blocks from which young mothers are tempted to throw themselves. I'd learned that the Romans invented, or discovered, concrete. They quarried a pink, sand-like material from Pozzuoli and combined it with silica and alumina to produce what became known as pozzolanic cement, which they then used in the theatre at Pompeii, in some of the arches of the Colosseum in Rome, and, adapted to include locally quarried materials, in Hadrian's Wall. The architectural historian Sir John Summerson has called concrete 'the Romans' greatest architectural legacy', which seems a bit unfair on, well, Rome – but it was more respectable than I thought. I splashed water on my face, took a deep breath, went back into the meeting.

'I don't want to lose the concrete,' I said mulishly.

It was difficult, though, to find anything else we might lose, because everything seemed essential. Eventually, reluctantly, we decided to dispense with the shutters and the hard landscaping around the outside of the house. The children's complicated built-in wardrobes could go, to be replaced, for the time being, with something from Ikea. We discovered that the rooflight over the stairs had been charged twice, which got rid of £2,000; we could fit out the utility room in cheaper materials and also simplify the timber shelving in the sitting room. Finally, turning Charlie's retractable rooflight over the bath into a non-retractable one would save another couple of thousand – which was hard, because for a long time, the house had been envisaged as an indoor/outdoor eating space with an open-air shower. Then the indoor/outdoor eating space went, and the open-air shower became a shower you could open to the air. And now even that was disappearing; something essential to our original conception had been lost. But it appeared that concrete had become even more essential.

Altogether, the cuts saved us roughly £20,000, which took us back into budget with roughly £10,000 contingency. That wasn't really enough, especially as there were still prices to be firmed up, but I didn't believe the costs anyway. The figures that Ramesh had submitted on his second, room-by-room, estimate had included prices for skirting boards, although there were no skirting boards anywhere in the house, and £400 for iron-mongery in the den, which only had one door. And why was he quoting £11,000 for painting, when the walls were concrete? There was something Wonderland-like about the figures. We had no idea how much a rooflight cost.

Suddenly, the design process stopped feeling like a marginally more sophisticated version of *Supermarket Sweep* – 'I'll have one of those' – and became hemmed in by sobriety and deadly seriousness. Now we had to take responsibility for what we were snatching off the shelves of Joyce and Ferhan's architectural emporium, because it was going to cost money. We were forced to cut anything that wasn't absolutely essential. That meant Charlie's outdoor bathing arrangements, which had sometimes seemed the main attraction, along with the boys' tree-house, which had never been designed and hadn't ever worked its way into the budget anyway. The fantasy part was over.

Varbud moved on to site on 19 November. We went along to meet them and were introduced first to Ramesh, immaculate in a blazer and roll-neck. 'And this is Mavji,' Joyce said (pronounced Maouji), 'who's going to be looking after the project.'

I shook hands with him and said warmly, 'You must be the site manager we've heard so much about.'

'Er, well, no, actually,' Joyce said. 'He's left.'

I stared at her.

'There was a falling out,' she said quietly. 'But,' she added, 'as a result we've got Mavji, which is *much* better. We hadn't even *hoped* to have him!'

Mavji, it appeared, would be running the job, although sometimes from the workshop, which didn't quite seem to make him a site manager. He had played with Ramesh as a boy in Kenya. Still, I liked the look of him: tall, with intelligent eyes behind his glasses. His manner was considered, almost intellectual, with an air of sizing things up, of preferring to think than rush about doing. He chain-smoked, narrowing his eyes through the haze that partly hid him from view, preferring to take a drag or flick some ash rather than commit himself. Both Joyce and Ferhan (obviously we didn't discuss this at the outset, only later in some hysterical moment when nothing seemed to be happening) thought he was sexy. And he was: mysterious and handsome, cool to the point of appearing only just to be ticking over. A few months later, he was banned from driving for a year for repeated speeding offences, which was funny, really, because I never saw him do anything else fast.

Mavji, then, was the man who was to set the pace for our

project. But we were also introduced to VM, who was going to be in charge of it 'on a day-to-day basis'. So he *was* a sort of site manager, I supposed.

VM was shorter and podgier than Mavji. He wore neat jeans and trainers, drove a silver Mercedes and had a lovely smile and mischievous eyes (once, Ferhan phoned him and asked, 'Is that VM with the beautiful eyes?' to which he replied, deadpan, 'Yes'). But the smile slid out only rarely; like Mavji, he resisted too much intimacy. This may have been in case we asked them to do something they weren't supposed to do, or somehow disrupted the chain of command, about which they had a strict company policy. But I felt at this first meeting, as I would many times afterwards, that they used their Indianness to close us out. (They had a habit of discussing the work in Gujarati. It used to drive Joyce mad when they did it in front of her and she'd tell them they were on no account ever to do it again.)

Ramesh explained now that they intended to spend the first week clearing the site and get the pilings in during the second. They should be able to get the ground beams finished in a fortnight, and then they'd either pour the slab (the floor, effectively) just before Christmas, or just after.

I was distracted before Christmas, as I always am. Ned's birthday falls in December, and there are the carol services and end-of-term performances, the buying of presents and sending them and tree-choosing and turkey-ordering and pudding- and mince pie-making and putting in appearances at parties and filing newspaper copy with a hangover. So I didn't really register until after Christmas that the ground beams, which Ramesh had explained on 19 November were two weeks away, still appeared to be two weeks away.

The builders had taken a fortnight off for Christmas as planned. (I later found out that this is when they traditionally built a bit more of Lux's kitchen.) But on 7 January, when they should have been back at work, Clinton Grobler called from Azman

Owens to say that no one from Varbud was on site this week because they were 'dealing with services' first.

Clinton was a South African architect, very precise, who had worked on the construction drawings for our project, in particular on getting services into the building, which demanded absolute accuracy, because once the concrete was in, it couldn't be hacked up again, and in this building there would be no visible ducting, no skirting boards or cornicing to hide things behind. But Clinton's attention to detail, so useful in many respects, could sometimes be slightly narcoleptic in conversation. It happened that he called when I was in the middle of some work and I didn't press him to explain. So it was only afterwards that I started thinking that I didn't know what 'dealing with services' meant, but a week sounded a long time to be doing it.

Charlie and I started wondering if we should have hired a project manager. We'd seen an episode of *Grand Designs* in which Kevin kept saying to the couple, 'Do you feel you should have hired a project manager?' and they agreed that they should have. Maybe that was our mistake as well? Maybe Azman Owens's interests weren't identical to ours, and they were running the project to suit themselves? (Though why they should have an interest in delaying the ground beams for nine weeks was a bit of a mystery.) The site looked chaotic: muddy, littered with blue plastic tape, polystyrene chunks and pink foam. It made me think of First World War trenches, or poisoned patches of ground in Eastern Europe, chaotic and hostile. We didn't understand what was going on, or not going on, and no one was telling us. We weren't entirely sure what a project manager did or how you went about finding one, but we felt we were missing out. Unfortunately, a project manager would presumably have been another person wanting a percentage.

Talking about a project manager was a way of not talking about Charlie's tax bill, which had finally arrived. We had done little to prepare for it, beyond worrying and clinging to a Micawberish belief that an unexpected large cheque would suddenly arrive on the mat one morning.

In December, Charlie had spoken to Cora, our accountant, about the possibility of staging the payments. 'Ooh!' she had said, sounding frightened. She promised to talk to the Inland Revenue, whose response was best described as extremely negative. The only thing Cora could suggest was that Charlie might write to them himself. So he did, proposing to pay half in January, followed by a series of payments in March, May, June and August (by which time he'd have had another bill, but he boldly proposed that he'd clear that as well. The house was scheduled to be finished at the end of June, meaning that we could sell the one in Hackney and use the capital.)

The tax office wrote back to say they permitted rescheduling only in the direst circumstances, i.e. bankruptcy. We felt bankrupt; effectively, we were bankrupt. But that wasn't good enough. A meeting was arranged for Charlie at his tax office in Euston for the first Monday in February.

That weekend, Charlie was preoccupied with what he was going to say. I was due to fly to New York, also on the Monday, to interview the ancient but kittenish founder of *Cosmopolitan*, Helen Gurley-Brown. I was preoccupied with leaving the family and whether she'd despise me for not having a manicure. We sniped at each other and wished we had a project manager.

On Monday morning, I spoke to Joyce before I left for the airport. She assured me the ground beams were two weeks away. That would be just the eleven weeks late, then.

Charlie drove down to Euston, parked the car, and left his bulky file of mortgage and bank statements on the roof while he fed the meter. The wind flipped the file open and tossed the papers into the traffic. They whirled into the path of the stop–start taxis, under the wheels of cars, up side streets where they flapped around trees and slid into the dogshit below. He ran after them, retrieving what he could. But some were whisked away or torn apart under the rumbling tyres of trucks.

He tried to recover his composure for his meeting but he'd felt like a criminal even before he'd lost half his paperwork.

Alison Charles was a fierce young woman of West African descent who towered over him and began by announcing that his schedule was out of the question. Charlie gave the little speech he'd been mentally rehearsing all weekend, explaining about how he believed in paying taxes, how he'd paid taxes for twenty years, always previously on time, about the house taking longer than expected, the two mortgages and the £50,000 in personal loans we'd had to take out to tide us through the autumn cash flow crisis . . .

Alison Charles's eyes began to roll. Her manner changed. She stopped treating Charlie like a delinquent and started treating him as a sort of hapless, misguided idiot. She thought for a bit, and then she came up with a schedule of her own, similar to his, except that everything would have to be paid off by May.

Charlie agreed because he had no choice. She said she'd put it to her superiors.

We waited and waited. She didn't phone, which felt like bad news, because by the end of the meeting she'd been on Charlie's side. Then he got a letter saying the proposal had been rejected. The Inland Revenue expected him to pay the whole amount in two weeks. If he didn't, they would issue distraint proceedings and come round and seize our property.

I looked around the house. I wasn't sure our property was worth that much. They'd have to take *everything*. I wondered how long it would take for the novelty of eating off the floor to wear off. Three days, I reckoned.

'Oh well,' I said, trying to look on the bright side, 'they'll probably leave our bed.' The base of our bed was broken; like so much of our life, getting a new one was in abeyance until the house was finished.

The letter included a slip explaining that the tax offices were about to close for refurbishment. They wouldn't open for a fortnight. This, rather luckily, meant that in reality we didn't have two weeks, we had four.

Towards the end of March we managed to pay another chunk. When the tax office reopened, Charlie sent a letter explaining that he'd now paid off three-quarters of the bill, and it didn't make sense for the Inland Revenue to take legal proceedings that would prevent him from paying the rest, especially on account of having four children to support, etc. He requested another meeting to discuss paying the outstanding amount.

We didn't hear anything for six weeks. Charlie claimed that he felt like the father in *The Railway Children*, about to be whisked off to prison leaving his family to fend for themselves (although, of course, there's a crucial difference in that the father in *The Railway Children* turned out to be innocent). Then someone called Ian Cameron called, several times, leaving what Charlie described as a series of 'spooky' answering-machine messages. In reality, these seemed to consist of the words 'Hello, this is Ian Cameron,' and a telephone number, but Charlie claimed they made him feel sick. We guessed that Ian Cameron was part of some tough, focused department of the Inland Revenue reserved for hard cases. He had a scarily efficient telephone manner and clearly wasn't intending to give up. Charlie finally steeled himself to call back. Ian Cameron wanted him to speak at a conference of the senior management of the Inland Revenue. Perhaps he could address them on the subject of innovation?

Eventually someone did get in touch, a man called Mark Saunders. And he turned out to be the Steve Symonds of the Inland Revenue, according to Charlie (this was the highest praise he could have given anyone at this point in our lives, implying flexibility, intelligence and grace). He suggested that Charlie should think carefully about what he could pay, which was the first time anyone had suggested compromise rather than removal of furniture. Perhaps, we thought tentatively, they were coming round to the idea of charging interest rather than locking us up. Charlie needed to do this over the weekend, he added: if he didn't call with his proposal on Monday or Tuesday, he'd have to wash his hands of us and our slapdash ways.

Early on Monday morning, Charlie called and said he'd pay half of what we still owed by the last week in May and the rest by the end of July. Mark Saunders suggested that the pair of them should get together in a month or two to discuss how we were going to manage the new bill that was coming in July.

A couple of years earlier, we'd had one mortgage and no credit card debt. We'd been the sort of people who limited our dealings in property to borrowing a bit extra on the mortgage to fit a new bathroom. Charlie, as he kept reminding me bitterly, had grown up in a family that regarded the acquisition of debt as evil. And now we were in a right mess. Still, Ramesh had produced a new schedule of works, which – paradoxically – showed us finishing on 3 June. This was three weeks earlier than originally planned, even though Varbud had asked for an extra two weeks for bad weather, which Joyce had granted, and even though we were, at this exact moment, at least thirteen weeks behind. We weren't going to point out the discrepancy. If we could sell the house in Hackney by midsummer, we would survive.

8

There was one cogent explanation as to why we still had a patch of nettles rather than a house; unfortunately, nobody bothered to give it to us.

Brian Eckersley had worked with Joyce and Ferhan on almost all their projects. He was tall, with fashionably very short hair and a hazily Northern accent. A qualified architect as well as a civil engineer, he could more than handle the maths (he'd originally planned to be a physicist), but the sums didn't interest him as much as the scope and ambition of the buildings. Brian had given up his job with a leading firm of structural engineers to build his own house on top of a dance floor behind a pub off Upper Street because at the time he could make more doing that than he did from his salary (unlike us, he'd bought his site for a few thousand pounds), and this independence gave him a cool, outsiderish demeanour. Architects kept seeking him out to work with them, finding him, as I did, a reassuring presence.

A few weeks before we went on site, Brian ordered a soil investigation. He asked a specialist firm to sink three boreholes, two to a depth of 6 metres and one to 12. He was concerned about possible desiccation of the ground caused by the roots of the ash tree, and this was probably what mainly determined his thinking about the positioning of the holes.

The results appeared to show that there was clay from about 5 metres down, which was pretty much what you would have expected in London. One hole suggested that there might be some gravel, a narrow layer perhaps, nothing unusual.

When the piling company arrived on site in late November, they hit a layer of gravel that went on and on. The site wasn't big. But it was big enough for the ground conditions to change abruptly halfway across.

The plan had been to sink something called augered piles, which involves a drill boring into the ground like a corkscrew, bringing out a plug of clay, and leaving behind a neat hole. Then you put in your reinforcement, slop in the concrete, and you've got a piling. But you can't do that with gravel because it keeps falling back in.

The piling company decided to try another method, known as continuous-flight augered piling, in which the concrete is poured down the middle of the drill simultaneously as it bores into the ground. But this requires a bigger rig. And when the piling contractors tried to get it on to the site, it wouldn't go down the lane.

There could have been an ugly dispute about exactly whose fault this was. Fortunately, Brian managed to redesign the foundations so that there were bigger piles and more of them, which meant that they could rest on top of the gravel. This in turn required some redesigning of the ground beams.

Nobody told us any of this. Presumably at architecture school students are strongly advised to avoid giving clients any information that might encourage them to make more of a nuisance of themselves than they already are. In the RIBA bookshop not long ago, I found a work of fiction from the 1930s called *The Honeywood File*, an epistolary novel supposedly written by architect, client and contractor. The client, Sir Leslie Brash, is always making unhelpful suggestions ('an influential acquaintance of mine is financially interested in a new novelty super-paint . . .'). It's a very dated and (especially if you are a client) not funny book: I expect it stays in print so that in architecture school they can hand it out to students in the first term.

Inevitably, things are going to go wrong on a project as complicated as the building of a unique house. I suppose architects and contractors don't want to be for ever apologizing when they could more usefully be putting them right. All the same, Charlie and I found the sense of being managed unnerving.

Not long before we started on site, we happened to meet a woman at a party who'd recently been a client of Joyce and Ferhan's. Did she have any advice, I asked brightly (I was not yet in my depressed phase). 'Get to know your builders,' she said, 'and go to all the site meetings.'

It was difficult to get to know the builders. Whenever I turned up they seemed to be in their little hut eating curry and khichri, the rice and lentil mixture that is the staple diet back home. Sometimes they'd be drinking from their flasks of Indian tea, which is made from the same ingredients as British tea – leaves, water, milk – but put together in a saucepan cold, brought to the boil and simmered for two to four minutes. It tastes of tannin, makes your head buzz and takes the enamel off your teeth.

The labourers often didn't speak English anyway, or they chose not to. Mavji or VM would always answer questions, and Mavji, in particular, would walk me round the site and explain what was happening (though I always felt slightly as though I was under escort). But the idea of getting to know them was futile. I wasn't there often enough and when I was, they seemed to want me to go away again. Joyce could shout at them, especially when she asked a question and VM passed it on in Gujarati (though she didn't shout at Ramesh, who didn't respond to shouting, she said), but I was in a more distant relationship with them, and they carefully deployed their exoticism to keep it that way.

Still, I could go to site meetings, and the first time we saw Joyce and Ferhan after Christmas, I announced that this was what I wanted to do. I was feeling groggy and bewildered because I'd been up since 5.30, seeing Hen off at the airport for her six-month gap year trip to South America. Before anyone else in the room had been up, I'd been weeping at Heathrow, and I was still struggling to get my head around the idea of her being gone. Joyce, by contrast, was her usual energetic and positive self, at least until I made my site meeting bid.

'But there's no point,' she said. 'Site meetings are boring and long-winded and you wouldn't understand them.' She meant we'd get in the way.

'All the same,' I said, 'I'd like to come. Otherwise I feel out of the loop.'

'It will make everything take so much longer,' said Ferhan, meaning we'd get in the way.

Charlie tried a different tack, explaining that those of his projects where the clients were most engaged tended to be the ones that finished on time. He was reasonable, Joyce and Ferhan found more excuses, I was tired and incoherent. 'OK, we promise not to ask any questions!' I said brightly at one point, which slightly undermined the point of going.

Eventually, we compromised an attendance at site meetings once a month, starting with the next one. I turned up at the site at 9.30 on a murky Thursday morning a few days later to find Mavji mooching about behind his fag and Joyce doing her best to ginger everyone up by complaining that the fax machine was too old and she wasn't having any useless old equipment on this site. The builders had put hoarding around the tree to protect its roots, set up a portable toilet and erected a little hut for their drawings and khichri eating; they were awaiting delivery of a container in which they would keep the fax, the phone, the plans and, in wet weather, themselves, though, to me, the site didn't look big enough to take one.

There was quite a lot of discussion about drains: we'd had to bring Dyno Rod in to flush out the mains, which were full of builders' rubble. As far as I could understand, we had to dig a run to the mains, and there was some dispute about where it should go. Joyce was right that there was an enormous amount of technical stuff that I didn't understand, but curiously, I found those parts the most comforting. I could let the impenetrable conversation about sealants and insulating materials drift around me and feel that something was happening.

On subsequent weeks, we had to rely on the minutes, even though they glossed the discussion so thickly that you could scarcely tell who'd been there. 'Service runs to be dug after completion of ground beams to house to prevent collapsing of trenches' sounded like a sensible decision. But what were these collapsing trenches? Did trenches usually collapse, and if not, why were ours doing it?

'Reinforcing cages are still in process of being put in position for casting of remaining ground beams,' Joyce wrote in March, where the 'still' probably covered some vitriolic argument about why everything was taking so long. 'The block has arrived for garage wall adjacent to neighbours' wall,' reported the same set of minutes. 'The pile cap of the original was in fact different from the drawings SE [structural engineer] had referred to and so the setting out was approximately 300mm off those shown in the drawings. It was agreed that the wall should be located as shown on the drawings and that the garage would become 300mm wider internally. The blocks will be laid tomorrow.' This came closer than usual to admitting error, but you still couldn't work out what had actually happened. And by the time we received the minutes, often a couple of weeks later, things had moved on, the problem had been solved or skirted around, and everyone was exercised by something else.

The tussle over site meetings was still rankling at the end of April when Hugo and Sue came over from Ireland. They'd bought a large, dilapidated Georgian house, south of Dublin, close to where Hugo had grown up, for which they were concocting plans for pared-down interiors. We were due, on one of their first trips back to London, to meet in Soho for dinner; Hugo was late because he'd been to see the site with Joyce.

'I've been trying to think all the way here in the taxi what I was going to say to you,' he told me, as he sat down.

'Oh, blimey.'

'You're a long way behind. I think you should be thinking in terms of September rather than the summer.'

'Mmm, I know that.'

'But it's going to be fabulous.'

How could he tell? All you could see were some reinforcing rods sticking out of the ground.

'I'm worried it's going to be too small,' I said. 'It doesn't look big enough to be a whole house. I think it's smaller than the house we're living in now.'

'That's why you employ architects: to make all the space work for you. You won't be falling over pushchairs in this house.'

'That, Hugo, is because we won't be able to take them with us. There won't be room.'

'I feel emotionally implicated in it,' he said, by which I think he meant guilty.

And, sure enough, much later he told me he had been horrified by how small the building (or lack of) seemed to be.

'Whenever I tell people I'm building a house,' I said morosely, 'they say "Oh, how fantastic!" I've hardly met anyone who isn't envious. They assume it must make you feel powerful and in control. And in fact, it's the opposite.' I told him about the tussle over the site meetings.

'You should insist,' he said, as if it were that easy. 'You should say to Joyce, "OK, we've done it your way. Now we're months behind schedule and we want to attend." And at the beginning of each week you should ask for a list of things that are supposed to happen in the next seven days, and if they don't, you need to know why not.'

The following afternoon, I telephoned Joyce and tried this. She remained extremely resistant. We'd slow things down, our presence wouldn't be helpful. Altogether it was a really bad idea.

'Either we come,' I said, 'or we hire a project manager.'

I'd just spent two hours on the internet trying to find out

about project managers – how you acquired one, whether there was some kind of professional project managers' organization – and drawn a blank. If Joyce had retorted, 'OK, do as you like,' I'd probably have to ask her how to go about it. But she didn't. She backed down.

I couldn't go to the next site meeting anyway, because I had to take Harry to look round a new school. And the one after that, I forgot.

On Boxing Day 2000, I'd dragged the family to look at our site; on Boxing Day 2001, we went to look at Elaine's. Clearly, her house was going to be very large, on account of its really being two houses. She was going to have rooms she didn't know what to do with. 'And this will be the World Cup Viewing Room,' she'd say as she walked you through the rubble.

More space was what *we'd* wanted; we were the ones with all the children. But now that we'd cleared the weeds, the apple tree, the garden roller and the abandoned chair, the patch of ground seemed smaller. Now that you could see clearly from one end to the other, it was obvious it wasn't far.

Varbud pegged out the footprint of the house in the earth. It appeared to be about as big as their container. Elaine took to calling it Tiny Towers, which was doubly aggravating as there weren't going to be any towers. It didn't cover much ground, and it wasn't going up very far.

'We should have built a four-storey gothic house,' Freddie said.

'It'll look bigger when it's finished,' I replied unconvincingly.

'Oh yeah? Will my bedroom be bigger?'

'I don't know. Maybe.'

'No, it won't. You know it won't. Honestly, what's the *point* of building a house if I end up with a bedroom the same size as the one I've got now?'

The point, Harry could have told him, was to have a tree-house.

He was standing frowning up at the hoarding round the trunk now, evidently trying to envisage it. (No one had told him it wasn't in the budget. I suppose we were still hoping vaguely that the money for it would be found, although I don't know from where, since we were having difficulty finding the money to stay out of prison.) 'I think,' he said pensively, 'that perhaps I'm also going to need a telephone.'

Hen had also come with us that Boxing Day, although her interest in the house-building project was intermittent – partly, and inevitably, as a consequence of her not being there a lot of the time, partly as a side-effect of her shuttling childhood, which had left her with a self-contained sense of herself, not linked to any particular location. She'd sought security instead in a network of relationships – friends as well as family – that she'd laced together underneath herself like a safety net. Even so – more than ever, in fact, now that she'd gone away – I was aware of how much I wanted this house to be somewhere about which she didn't feel ambivalent, in which her feelings weren't fragile, and to which she wanted, uncomplicatedly, to return.

Freddie, on the other hand, had always been instinctively domestic. His childhood had left him strongly attached to home; he liked being in, surrounded by his family. But while he was more interested in the project, he was also more disparaging. He didn't seem to trust us to provide for him in what he regarded as our folly.

One Sunday morning (this was not long after I decided to stop arguing with him and resorted instead to sighing through my teeth, implying he was being exasperating and unhelpful) we stopped at the site on the way to somewhere else. The younger boys gambolled about on the edge of slippery trenches with spikes at the bottom and skipped dangerously over protruding reinforcements while I teetered after them in my high-heeled ankle boots and tried not to think about them falling forwards and impaling themselves. Freddie got out of the car, took one disgusted look at the plot, and got back in again.

He might, I thought, be right about its being too small. But he was wrong that we had been negligent of him. The house was in part an attempt to draw breath, to establish ourselves properly at last, to give everyone their equal weight and place, to be legitimate. And that was about him as much as anyone. It was an attempt to make up for the times that his PE kit was in the wrong house and that, what with all the shuttling, he'd forgotten to do his homework and now it was too late; to recompense with a proper family home and, at the same time, to put it all behind us.

I knew he couldn't put it all behind him, really, any more than Hen could, that the effects of their peripatetic childhoods were deep inside them, but it was galling that he didn't trust that I was *trying*. I could still taste the memories of children skipping off down the front steps, knowing that I had to turn and go back inside and wouldn't know what to do with myself, that the point had drained out of the day. I could remember waking up on a Saturday morning, listening for them and realizing they weren't there. I could still feel the frustration of buying new socks only for one immediately to be left somewhere else, of trying to locate a missing jumper that might have been under a bed on the other side of London, or possibly in Redcar, where they probably wouldn't be going for another month, and how could anyone be expected to remember to look for it then?

Hen's bewilderment was a deep current that didn't surface properly until adolescence, when it swept her away on a tide of exasperation at what she scathingly referred to as her New Labour New Parents childhood. But Freddie's was ever present. For much of his childhood, it was as if the things that might have given him ballast – the book he was reading, his homework folder, his pencil case, his jumpers (they needed twice as many clothes as other children, what with the ones you wanted always being somewhere else) – were whirling centrifugally around him, just out of his reach. I looked on, feeling helpless, thinking that good parenting was largely a matter of being in control or, at least, of giving your children the illusion that you were in control, and infuriated that I wasn't able to make the simple things work for him. Since I couldn't go and look under the bed in Redcar myself, I nagged at the other household, earning myself a reputation as a fusspot, my focus on the trivial a mean-spirited war of attrition against their larger, more oxygenated ways of sitting round tables discussing the future of this or that cabinet minister. I hated feeling unable to help my small, vulnerable boy, hated how vindictive it made me feel (there was something in the war of attrition theory, no doubt), hated the knowledge that I had brought this situation about.

The house, at some level, was an attempt to prove that I *could* be in control, if only because we would be starting from zero, from a hole in the ground. I wanted to build somewhere we all belonged, felt comfortable, a house from which we were free to go at times and under conditions of our own choosing, and to which, consequently, we were relieved to return.

I took to visiting the site in the late afternoon, after 4 p.m., by which time the builders had left, and there was no danger of surprising them in some disconcerting tableau. (I once intruded upon two labourers peaceably combing the earth between the trenches into neat patterns.) Later in the day, I could stroll around without disturbing anybody, without feeling that they felt the need to talk to me and to fend me off.

The broken wall that had originally bordered the site had gone, replaced with metal fencing panels tied together with plastic rope, or clipped at the top with U-shaped pieces of reinforcing steel. It was easy to break in, and I'd scramble across the rutted ground, between bales of the pink insulating material destined for the foundations, and pace between the stands of reinforcement that stuck up to show where the walls would be. I'd try to imagine myself in these rooms, having a conversation with Charlie in the kitchen, doing a jigsaw with Ned on the floor of the den.

The trouble was that the more the house assumed a shape, the more difficult it was to envisage. Short of erecting a full-size mock-up on the site, it was impossible to grasp exactly what we were building. Charlie said he tended to focus on one or two images, like the concrete wall behind the stairs. I didn't know which wall he was talking about. I wasn't aware that there *was* a concrete wall behind the stairs. Or, he said, a tranquil bedroom that wasn't full of his work. He was finishing his book, and there must have been twenty or thirty piles of books in our bedroom, none of which could be touched because they were all carefully ordered. Joyce and Ferhan thought it was dreadful that I had to climb over obscure works on futurism to get to my T-shirts, but actually, it was worse for Charlie. He hardly left the room.

Charlie reported back from the first site meeting I missed (the one I hadn't actually forgotten). Joyce arrived late: she had bronchitis and shouldn't really have been there at all. Clinton arrived early and picked a fight with Ramesh about the slow rate of progress, to which Ramesh coolly replied that he was used to working cooperatively rather than confrontationally and on a job like this, there had to be trust. Then he left. Clinton said to Charlie, 'So we'll play it his way for the time being,' as if we had any choice.

There were a couple of desultory labourers, working with drills (when Charlie demonstrated this, he did it one-handed). Mavji was 'semi-apologetic' and explained that the trenches were collapsing because of the rain.

Still, there were four concrete test panels, which were what had made Joyce get out of bed. And David Bennett had come in from Essex. 'What was he like?' I asked excitedly. Charlie said, 'Exceptionally interested in concrete.'

I had believed that these panels were being cast to test the colours created by using different types of aggregate. For some reason I persisted in this belief even when Charlie reported that the discussion had entirely focused on the length of time they'd remained in the formwork: the first panel had been left in for one day, the second for two, and so on. The longer they cured, the darker they became.

Charlie had asked Mavji what would be the effect on the schedule of leaving them for four days; Mavji, he said, 'seemed nonplussed', from which Charlie deduced that there was no provision on the timings for leaving the walls in the formwork for four days. Another interpretation might have been that Mavji knew the schedule was a fiction. Anyway, the point seemed irrelevant to me, since we didn't want dark, four-day concrete, we wanted the palest walls possible.

I resolved to look at the panels as soon as I could, but in the meantime we were leaving that afternoon for Dublin to see Hugo and Sue, who were now camping in their (very large) eighteenth-century Georgian house in Dun Laoghaire and mulling over architects' plans to strip out and streamline their interiors. (Sleek, Hugo said, hadn't really reached middle-class housing in Dublin yet. Perhaps, I thought mutinously, that was because sleek was impractical. I doubted my ability ever to live other than congestedly, unless in a mansion, and we clearly weren't getting one of those.)

While we were in Dublin, we visited the city's latest architectural addition, the extension to the National Gallery of Ireland. 'Hah, that's interesting!' Charlie said knowledgeably, looking at the concrete. 'Chamfered edges! We're going to have those. They're to stop the formwork damaging the panels when you peel it off.'

I had no idea what he was talking about. I was mainly interested in the colour, which was white.

But when I went to see our test panels, immediately after the weekend, they were grey. Some were a deeper grey than others, but they were all, unmistakably and incontrovertibly, rather grubby and dark. 'Well, those won't do *at all*,' I muttered irritably. 'I'll have to call Joyce.' But I couldn't get hold of her that afternoon, and the following morning, the minutes of the site meeting arrived. 'Concrete panels approved,' I read, 'by AOA, concrete consultant and client.'

I stomped upstairs to see Charlie. 'Is this true?' I waved the piece of paper at him. 'Did you approve these panels?'

'Huh?' He looked up from some book about globalization.

'Did you approve them?'

'Er, no . . . Well, I suppose I might have said, "They look fine." The lighter two.'

'Oh, this is ridiculous! How can they do this? How can they say, "approved by client" when only half the client was there?'

I trailed back downstairs and telephoned Joyce. 'Those panels were never approved!' I protested. 'I wanted beige concrete!'

Beige, Joyce explained patiently, as though to someone of limited comprehension, was not possible. David Bennett had spoken to a number of ready-mix companies and decided to go with one. This was the colour they did.

'So I can have any colour so long as it's grey? What was the point of all that stuff in David Bennett's book about different aggregates? Ground-granulated blast-furnace slag cement blah blah? And how come other people can have pale concrete? The concrete in the National Gallery of Ireland is nearly white!'

'That's a much bigger project,' Joyce said patiently. 'If we'd wanted a special colour, we'd have had to get the concrete company to clean out their vats every time. It would be very expensive, plus they might not do it properly and then we'd end up with different colours.'

Even I could see that different colours would be worse than grey.

'So how did Konditor and Cook get their concrete?' They'd recently opened a new, Azman Owens-designed shop on Gray's Inn Road, with a handsome concrete panel beside the door and a concrete counter, both pale.

'They mixed it on site. The quantities were smaller.'

We had the wrong size of job. It was both too big and too small. If we'd been bigger, we could have had a batching plant on site; if we'd been smaller, we could have made up the concrete in the quantities we needed. But we were planning to pour once or twice a week for two or three months. We couldn't expect a ready-mix company to set a vat aside for us for all that time.

'So you are saying,' I asked futilely, 'that it's absolutely impossible to look at different colours?'

'It's not impossible, no. Nothing's impossible; but there would be time and cost implications.'

'Meaning?'

'The house wouldn't be finished until Christmas. And we're paying £62 a cubic metre for this concrete. A different colour would cost more than a hundred.'

White concrete, David Bennett told me later, would actually have cost a minimum £400 per cubic metre. While I'd been drifting along, imagining I could pick my colour, rather as if off a paint chart, David had interviewed a number of ready-mix concrete companies. He'd weighed up the seniority and calibre of the people they put forward to talk to him, the prices they were quoting, the ease of access to the site (it was important not to be too far away; you didn't want the mixture swilling around in the drum for too long in case it affected the consistency). And he'd decided to go with a company called Hanson because they met all the criteria, he'd worked with them before, and they'd undertaken to send a technician with each load.

Much later, David told me that there might have been one

affordable way of getting a paler concrete. We could have had some blast-furnace slag in place of a proportion of the Portland cement, which is an option that a number of ready-mix companies offer, not least because the blast-furnace slag is 10 per cent cheaper. But, he said, 'the cement batcher could have made a mistake in the blending and if we were slightly out we could have ended up with a different colour on that load. Hanson were offering us good quality controls and I didn't want to make it more difficult than it needed to be.'

This at least, was information. It was more than I'd got so far. I couldn't believe no one had involved me in the decision, knowing how agitated I was about pale concrete; nor that once it was made, they hadn't at least explained it to me.

'Besides,' Joyce said wearily, 'the concrete is supposed to look like concrete. I don't actually like the concrete in the National Gallery of Ireland. It looks like plaster.'

'Well,' I said bad-temperedly, 'I *do*.'

The builders wanted to move Tom Tasou's gates, since they were still blocking access to our now pegged-out house. Joyce suggested that we should put manual gates up temporarily, since to reinstate the electronic gates and get them working properly would cost upwards of £3,000. We weren't yet in a position to hang them in their final place – firstly, because they'd be in the way, and secondly, because they were meant to hang off a concrete wall that we still hadn't built.

Somewhat surprisingly, Tom Tasou agreed to this proposal, perhaps because he'd finally sold the house at the bottom of the lane, perhaps because the gates had only ever been a means to an end. At least, he agreed to it initially. A week after the manual gates had gone up he rang again. Every single one of the tenants had complained, he said. They wanted compensation. We needed to re-install the electronic gates and we needed to do it soon. Meanwhile, we probably needed to 'do some PR'.

Joyce didn't want to distract the builders from the preparatory and piling work and she was worried, I think, that we could end up having to take the gates down again more than once. So we composed an apologetic letter to our future neighbours explaining that we intended to get the electronic gates working as soon as the pilings were complete and the front wall was cast. We couldn't say exactly when this would be, but we thought a matter of weeks.

Days later, we received a solicitor's letter from the owners of one of the three mews houses, a Mr and Mrs Gregg. Their tenants had threatened to quit and they wanted the gates replaced *forthwith*. This was written in solicitor-speak, i.e. as if they were about to beat us up.

We asked Ramesh to quote for putting the electronic gates back, not in their final position, but as far to the other side of the lane as possible, so they wouldn't be in the way. A month later, he still hadn't come back to us, but we'd had another solicitor's letter. The Greggs claimed their tenants had agreed to stay only if their rent was reduced by £50 a week, backdated two months, and they considered us liable.

Bridget, our solicitor, said she couldn't see on what basis they could litigate (although she admitted she wasn't a litigation lawyer and they might well find something). The other tenants didn't seem to be finding the situation insufferable, she said, and there was no indication that the Greggs couldn't re-let and maybe get even more money.

On the other hand, she thought the loss of security and the electronic buzzer system probably did constitute more than ordinary nuisance. I agreed with her here, even though I thought the gates were ugly and pointless – why would you want to keep us out? – plus politically incorrect, in the sense that if everyone in London decided to live in gated communities the city would cease to function. But since there had been gates when the tenants moved in, it seemed to me they should probably have them back.

If we did decide to settle, Bridget warned, the Greggs couldn't arbitrarily decide the amount of compensation. And we should be aware that we might set a precedent and end up compensating everyone in the lane.

Even as Bridget and I were having this conversation, the fax machine clicked, whirred and churned out the quote from Varbud: £4,128, to include automation and intercom. Eagle Security, the company that supplied the technology, would need two to three weeks to schedule the work. I called Varbud and asked Sri, who organized the work, to get it moving as quickly as possible. We also phoned the Greggs (who were much nicer in person than in their alarming solicitor's letters) and explained that things were finally happening. The automated gates went back up, on temporary fixings, in the middle of April. Two weeks later, all the buzzers broke again.

In the middle of all this to-ing and fro-ing about the gates, Joyce took us aside and explained that whereas in most houses the load-bearing walls didn't take long or cost much to build, in ours, because of the concrete, they constituted a large part of the expense. Under the terms of our self-build mortgage, we could only draw down the next payment to Varbud, of £150,000, when the first-floor slab (i.e. the upstairs floor) was finished. But Varbud had already been on site three months. They needed £15,000 now, this week, and another £60,000 in the next fortnight.

Joyce suggested that she should write a letter for us to present to the Woolwich, so that we could see if they would give us any flexibility. If we'd got the original self-build mortgage from Barclays, the money would have come to us in seven stages, but the Woolwich's version allowed for the money to be released in five parts, which didn't happen to relate particularly well to the order in which we were doing things. (We'd had a slug of money upfront, then the subsequent stages were: first-floor slab; dry – which meant the roof on and the windows in; first finish – i.e. electrics and plumbing; and second finish – plastering and painting.) They ought, Joyce said optimistically, to be able to see that their schedule was to some extent arbitrary, and that it wasn't working here.

But what if they didn't? If we couldn't pay them, would the builders simply stop? And then how would we ever get the money?

The people at the Woolwich were intransigent. We could have the money when the first-floor slab was poured (as it said in the mortgage agreement, they pointed out testily) and not in dribs and drabs before.

Steve Symonds was not at work. No one seemed to know when he'd be back, only that it wouldn't be immediately. He was unwell, was all they'd say. This was a disaster: we'd got used to him soft shoe-shuffling his way through 'the Barclays environment' on our behalf, sorting things out. Charlie spoke to someone else at Barclays about the possibility of a bridging loan.

Only a couple of months earlier, I'd rashly announced in a meeting with Joyce that I'd been thinking over whether the house had had any affect on Charlie's and my relationship, and (I was probably rather smug here) that I had decided it hadn't. We hadn't disagreed about anything (of course, this was easy: we'd just done what Joyce and Ferhan told us); we hadn't discovered that we had radically different ideas about our future. We still weren't disagreeing about the house, but now the strain of the money was making us quarrel about all kinds of other small things. Charlie was about to go to Australia, where he was due to give some talks and no doubt meet interesting people. In the few days before he left, we had two fights, the second over the lunch that was supposed to make up for the first.

Charlie was trying to sort out the bridging finance before he left the country for three weeks; I was fighting off persistent low-level, nagging concern about Hen, who was billeted in a suburb of Guadalajara, sharing a bed with a girl she didn't know or, it seemed, like. When I muttered about the appalling organization of her gap year company, Geoff, a friend of mine, said robustly: 'It's good for her. Most of the world is like a suburb of Guadalajara,' which may have been true but somehow wasn't much of a consolation. She bunked off her job, which wasn't really a job anyway, and took a week's side trip to the coast where, she said, there were too many dismal hippies. She met a couple of boys called John and Patrick, with whom she planned to stay at a ranch with some Mexican men and then go on into the jungle. There was a long stretch when I didn't hear from her and it occurred to me that I didn't know John and Patrick's surnames, or if they were, as I believed, students of Spanish at UCL, or if I'd merely made that up in a hopeful moment, or what their itinerary was. Or even where Mexico is, very accurately.

The day before Charlie left for Queensland, we dashed to a branch of Barclays in the City to sign the papers for something called an executive loan. We'd be able to pay this back, we trusted, when the first tranche of the mortgage came through. It would cost £800, and we swallowed and pretended it didn't matter, or it wasn't happening or something, because there was no point in thinking about that now. We were on a ferry in pitching seas, too far out to make it safer to go back than to go on.

I was fed up with feeling like a small disreputable state whose lines of credit might be withdrawn at any time. The persistent worry about money made me feel exposed, as if I might be caught out – not just at the cashpoint or the supermarket till, but everywhere. When the car broke down in a village in Hampshire and I had to wait in the dark for someone to come out from a garage in Portsmouth to rescue me, I rang Charlie in Australia, even though for him it was the middle of the night. Being stranded in the dark just seemed to confirm my vulnerability: here I was, stuck on the edge of a field with a useless car and, probably, a useless credit card. It was suddenly rather annoying that he was asleep.

The week before Easter, I crossed the road to look at Elaine's building project, which was nearly finished. There were ten people working there. Then I drove over to Highbury to look at ours, which had barely started. There were two people working there.

A day later, Joyce copied us into a fax that she was sending Ramesh. 'We are concerned that progress on site has been slow and are aware the construction programme has fallen behind,' she wrote severely.

> You have verbally informed us of several delays. However, in accordance with Clause 2.3 (IFC 98) if the progress of the Works has been delayed by any event, you must inform us in writing of the cause of the delay as soon as these matters become apparent. Would you please state which items outlined in Clause 2.4 that you consider apply in this instance, and forward such information that you consider will reasonably enable us to form an opinion and determine a fair and reasonable extension of time.
>
> Then a revised construction programme must be issued promptly to bring the Works in line with the agreed completion date.

I was due at a site meeting the following day, and I suspected the fax had been written partly to pre-empt my dismay at the lack of progress. In this sense it worked, because I arrived with pitifully low expectations. They were not exceeded. Ramesh put in an appearance and assured us that Varbud were doing no other work over the next month, in order to devote all their energies and resources to the project.

'I don't understand why it's taking so long,' I said pathetically, since Joyce obviously didn't either.

'It was very wet in February,' she offered hopefully.

'OK, let's assume it rained all twenty-eight days in February. [It hadn't.] We're four months behind. How are we supposed to account for the other three?'

She had to admit it was difficult. The rain argument didn't do much for me anyway: what had they expected in February in Britain? And since then, we'd had unseasonably warm weather: hotter than the Mediterranean, they kept saying on the radio. It seemed rather unfair that you lost time for rain but didn't get any credit for an off-season heatwave.

Two weeks later, we still hadn't seen the revised schedule. Joyce was forced to send another strict fax:

> Further to our letter dated 20 March 2002, regarding progress, a revised programme and a completion date, we have received no reply from you. Your immediate attention is required in this matter.
>
> We are growing increasingly concerned that the programme is drifting and now is the time to rectify the matter. Our clients need to schedule their move as well as the sale of their house.
>
> Tomorrow morning the client and myself will be on site for our weekly site meeting and we expect a revised programme to be presented. Included in the programme is the necessity to provide a detailed proposal as to how the concrete walls will be scheduled. The considered programming of the walls will allow you to provide a realistic programme and push everyone to remain on schedule. We remind you, as advised by David Bennett, the first wall to be poured should be a small and manageable pour. Please take this into consideration.
>
> We await your response.

Ramesh turned up at the following day's site meeting with a schedule that took us up to 28 July. Charlie and I were relieved. My private deadline was September. I didn't particularly want to move in the summer holidays anyway, and this made September seem still feasible. We were pretty certain we'd sold our house: some friends of friends (she was a paediatrician, he was an academic: they lived round the corner in a slightly smaller house, with three teenage daughters) had approached us back at Christmas and asked for first refusal at the asking price. Since then we'd had a valuation, which they'd accepted was fair. We had agreed to sell to them at the price, and they'd promised to hang on for us.

The slab, which Ramesh had told us back in November would be poured possibly the week before, possibly the week after Christmas, was actually poured in the week after Easter. It was a mystery why it had taken so long. Everybody seemed to blame Varbud; Mavji looked into the middle distance a lot and shook his head gravely. It wasn't even as if, having waited so long for it, it was interesting. It was ineffably dull, as bits of construction go (which is often quite dull): a flat bit of concrete with little groves of reinforcement sticking up out of it. Still, we had Ramesh's new schedule, which promised that things were going to move much faster now.

9

I was becoming a concrete bore. With all the zeal of the convert, I started buying books. I went to an exhibition at the RIBA. It became my ambition to meet David Bennett.

I could have told you all sorts of things about concrete: that it is a composite material, consisting of sand, aggregate (usually stone chippings or gravel), cement and water; that its name is derived from the Latin *concretus*, meaning grown together; that the architectural critic Peter Rayner Banham described it as 'a messy soup of suspended dusts, grits and sump aggregates, mixed and poured under conditions subject to the weather and human fallibility'. I could have told you that the Victorians were suspicious of concrete, on account of its plasticity, thinking of it as an ersatz material – acceptable for bridges or docks, but not for 'proper' architecture.

I have some sympathy with this view, because concrete's versatility has always made it a difficult material to understand. It isn't so much one material, in fact, as a whole spectrum of them. By changing the type of aggregate, concrete can be made light enough to float on water or twice its usual density. It can be made totally impermeable, for use in giant dams, or porous enough for filter beds at sewage treatment plants. As a result, its personality is promiscuous: it is capable of being 'as smooth as a cashmere jacket or as rough as hell', according to the architect Piers Gough, who also claims that 'in South America, concrete is sexy, like the culture. In Britain and Northern Europe, the material is intellectual but cold.'

Following the initial, Roman, flurry of excitement about it, concrete went out of fashion for a long time, mainly because it needs reinforcing to be really strong, and effective methods of reinforcement weren't invented until the nineteenth century. The modernists made a fair amount of experimental use of it, but it wasn't until the second half of the twentieth century that it really took off, when it appeared to be the cheapest, most versatile, fastest option for the rebuilding of Europe's shattered cities. By 1965, there were 230 competing systems of factory-produced, site-assembled concrete components available in Britain. In a matter of a few decades, concrete had swept the world, either in the form of pre-cast panels that could be assembled like masonry, or in-situ concrete poured into more or less rough wooden moulds with the reinforcing rods left sticking out of the top, ready for whenever another storey became affordable. Concrete is now second only to water as the world's most heavily consumed substance: one ton per year for every person on the planet.

This welter of pre-cast and rough in-situ stuff was quite different, however, from the fair-faced poured concrete that we were contemplating, even though the concrete enthusiasts have a confusing tendency to exploit the material's multiple personality disorder to have it all ways. They harp on about what a marvellous cheap material it is, and also what a beautiful one. But on the whole, cheap concrete isn't beautiful and needs to be plastered over or faced with some other material in any setting where human beings must spend a lot of time. It is only where as much care is taken over the formwork joinery as if you were building a dining table (this, David Bennett said, was the standard he was looking for) that concrete becomes beautiful. The attention that has been paid, the care that has been lavished on it, is what makes it lovely.

The general confusion about what this chameleon-like substance stands for is aggravated by the fact that brutalist architects used concrete to mimic the roughness of cheap utilitarian construction – giant silos, grain elevators – for broadly artistic ends. And something of this imagery attached itself persistently to the material, so that concrete still conveys ideas of urban, industrial modernity and hard-headed socialist brutalism. Finally, though, in the early twenty-first century, people are becoming more alert to the multifarious possibilities of concrete, and in particular, its use as a quality material in high-end buildings. Tadao Ando, as ever, has been crucial here, but many of the world's most interesting architects – Herzog and de Meuron, Rem Koolhaas, Denton Corker Marshall, Zaha Hadid among them – are exploring its potential. Award-winning buildings keep turning out to be made of the stuff. It is almost exhaustingly fashionable.

Meanwhile, and almost certainly related to its renewed respectability, the marks of formwork and signs of construction that were once deemed embarrassing have become acceptable, like the grain in elm or the marks of the chisel on sculpture. (In pre-war modernist buildings, it was common to whitewash the concrete to disguise its flaws. Le Corbusier's Unité d'Habitation, built between 1945 and 1952, was shocking partly because it was one of the first buildings in which the raw concrete – and it was really quite raw – was on show.) Now, concrete is increasingly seen as an organic material, which means that the marks of its making are not seen as shameful signs of fakery, but admirable (as long, anyway, as there aren't too many of them, and everyone knows in advance more or less what marks they are going to get).

Our concrete would speak to us, I hoped, of the effort that went into it. In one of my concrete books, I found something that Bernardo Gomez-Pimienta of TEN Architects in Mexico City had said about the concrete house he built in a small village. No one locally knew how to build in concrete, so all the people and tools had to be brought out from Mexico City, two hours' drive away. He persisted, nevertheless. 'I chose to use concrete because it is a wonderful material,' he explained, '– soft, monolithic and very sensuous. It is at the same time a structural material and a finish, and it is apparent how it is made, so there is a handmade aspect to it. A bit like watercolour, it shows every mistake, and allows for no corrections. Concrete is simultaneously very industrial and very handmade; it has the solidity of stone and at the same time the poetry of once having been liquid.'

This was what I hoped for from our concrete: that it would be unlike any other, because conditioned by our site, our builders, the weather, the mistakes that were made along the way. Its mottling would tell stories and its smoothness be a sign of human effort and attention. That was the idea, anyway.

We were due to pour our first concrete on 12 April, Harry's seventh birthday, but the formwork wasn't finished on time. Formwork for fair-faced concrete is always tricky (as David Bennett's dining-table metaphor implied), not least because if the joints between the shuttering panels aren't properly sealed, you can end up with pebbly seams erupting along your smooth surface. We were making it even harder by deciding to do away with tie-bolt holes. These are the indentations left by the bolting on of the vertical, horizontal and ground-to-wall joists pinned to the outside of the wood to stop it simply giving way when the concrete is poured. (Each one of Tadao Ando's 180 × 90 panels has six holes in it.) Holes are a risk, however, because they offer water an opportunity to migrate and break up the surface. We would still need bolts in our formwork, but if we were clever, we could have them at the top and bottom only, where they would be hidden by floor and ceiling.

The pour was rescheduled for the 15th, which was infuriating, because I had to be out of the country, interviewing Peter Ustinov. (I was slightly mollified by some very good concrete at Geneva airport.) I agreed to meet Joyce at the site as soon as I got back the following day. Varbud had already gone when I arrived, and so had the formwork: there was just birdsong and a wall, rearing up out of the slab.

When Joyce and Ferhan had talked about an envelope of fair-faced concrete, I envisaged walls at either end of the house. And this thought had lodged in my head and not budged, even though Charlie had alerted me to the likelihood of a wall somewhere behind the stairs. If I gave this any thought (and I didn't, really) I probably thought that after all, this was an external wall as well, giving on to the slug garden. But the wall we had now, the one we'd built, was in the middle of the house, separating the kitchen from the sitting room. It was an internal wall. Apparently we were having rather more concrete than I'd thought.

There was nothing I could do about it now. Anyway, something else was disturbing me more. Now that we had one wall, the likelihood of getting a whole houseful of them on to the pitifully small slab looked extremely remote. I mooched grumpily around the area that would soon be the slug garden, wondering what on earth had possessed us to dedicate such a big, sunless space to the breeding of invertebrates.

But perhaps I was misunderstanding. When he was younger, Freddie had been diagnosed as dyspraxic, which is a bit like dyslexic except that you bump into things. (In fact, all these fashionable child syndromes, including ADHD and Aspergers, overlap suspiciously: there's a theory that which one you get depends on which type of professional – physiotherapist, educational psychologist, teacher – hands out the diagnosis.) Freddie and I now both think he didn't have a learning difficulty so much as a teaching difficulty, which is to say, he had dyspraxic traits (which also include a predilection for daydreaming and an active imagination), but if he'd been taught properly they needn't have bothered him that much academically. Whatever, we didn't have to look far to see where the traits came from. I also have untidy handwriting and when people say that something is 250 yards away, I have only the haziest sense of how far that might be. So I had to accept that I might not be seeing the space properly, in all its wide-open possibility. It was only this that stopped me from calling Joyce and bursting into tears.

I touched the wall tentatively: the four-panel width was as smooth as slate and surprisingly warm. There were pale ovals imprinted on the surface, where knots in the birch had been plugged, and delicate horizontal tracings where the shuttering had been peeled away. Across the middle there was a stormcloud of dark mottling, with the haziness of a Turner skyscape.

I was impressed by its solidity and forcefulness. ('OK,' it seemed to be saying, 'you wanted a house; bet you didn't think you were getting *this*.') The evening sun was shining on the wall through the shifting leaves of the ash tree, shade and light skittering across its surface. I found it surprisingly unforbidding for such a big, dark, implacable thing, which I think was to do with its sensuous smoothness, its mysterious mottlings and muted, peaceful variations.

It might, of course, still develop the ugly streaks I'd been fearing all along. Did these generally happen straight away, or later, as the concrete dried out (or failed to)?

According to Stewart Brand, who among other things is the author of an excellent book about the adaptability of construction, *How Buildings Learn*, 'Concrete is subject to deterioration problems such as (alphabetically) blistering, chipping, coving, cracking, crazing, delamination, detachment, efflorescence, erosion, exfoliation, flaking, friability, peeling, pitting, rising damp, salt fretting, spalling, subflorescence, sugaring, surface crust, weathering . . .'

There was plenty to worry about.

Six days later, I finally got to see the concrete being poured. It should have been five days later but, according to Joyce, Varbud 'forgot to order the crane'.

'Isn't that, er, a bit incompetent?' I asked.

'Don't get me started. They need a woman to organize them.'

'What about Lux?'

'Oh, *that* side of the business runs really smoothly.' That was the money collection side, which wasn't necessarily what I wanted to hear. 'Still, they're very caring.'

The concrete mixer was due at 11 a.m., David Bennett having decided that this was the optimum time to drive from Stratford to Highbury. (He'd made them do a trial run, to ensure the concrete would be sloppy when it arrived.) The crane was already waiting at the bottom of the lane when I arrived. I parked the car on the forecourt of the workshop opposite, avoiding the severed Barbie-style legs strewn about and the one-armed naked women propped against the wall outside the mannequin repair shop. A van was coming down the lane, presumably collecting some good-as-new mannequins, or dropping off some tatty and injured ones for repair; I sprinted up and asked the driver to get out of the way.

'But you have to!' I protested, when he declined. 'The concrete's like a cake! You have to get exactly the right mix of ingredients and bake it quickly.'

He stared at me. But a concrete lorry was heading towards him, followed by a technician in his Motorway Maintenance vehicle; wisely, he thought better of continuing to argue with me, and backed into some garages to allow them through.

As soon as the concrete lorry was in position, the chute came down from the back and the concrete gushed into a hopper: a thick gumbo of grit, nothing like the smooth cement of my wall. If you stood within 10 feet of it, bits flew up and stuck to your jumper: mucky specks of stone, coated in sticky grey gloop.

The crane swung the hopper over to the narrow slot at the top of the mould, where the builders were waiting on scaffolding to tip it in. Mavji and another of the senior men plunged large vibrating pokers in after it to extract the air. (Already, Mavji had perfected the performance of this highly skilled and crucial job while appearing to be more interested in smoking.) The hopper went back for a refill.

The whole thing had to be done in an hour because after that you couldn't rely on the consistency. The concrete at the bottom might start to set before the top was finished, leading to what David Bennett described as 'a Battenberg effect'. At the same time, only a certain depth of concrete could be poured at one time, or the vibrators couldn't get down to get the bubbles out, resulting in a pock-marked surface with bits of stony aggregate showing through. Pinholes (of which we had quite a lot on our first wall) were generally agreed to be OK – charming, even – but big holes were not. So, for an hour, the site was a commotion of shouting in Gujarati over the noise of the Asian radio station, the creaking of the concrete lorry's drum, the pebbly cascading of the concrete and the roar of the vibrators. And then it was over, and time for Indian tea.

It was at this point that I finally realized that I was in the presence of the great concrete expert. David Bennett was a broad man of medium height with improbable iron-grey hair balanced precariously on top of his scalp. He spoke in a low but oddly not unmusical monotone, without appearing to draw breath.

David Bennett told me that he 'fell in love with the idea of concrete' in Iraq, when he was working as a civil engineer on irrigation projects in the south-eastern desert before the first Gulf War. Aware that the canals he was constructing could have been a concrete eyesore, he wanted to make them something else, and discovered in the process that he had 'an affinity with the material, this natural understanding'. Back in Britain, he went to work for the Cement and Concrete Association as an all-round concrete proselytizer for seven years before setting himself up as an architectural consultant, a sort of human irrigation channel for Concrete Knowhow.

Now he was concerned about my response to the black stripe across the first wall.

'Oh, I like it,' I said shamelessly, the woman who'd wanted white concrete. 'It's lovely: like looking at a picture, or a cloud formation, or the ripples in a pool.'

He wasn't sure how it had happened: whether the prop in the middle had sprung out a bit when the concrete was poured in, or had been pushing so hard that the formwork absorbed more water around it. 'But that's the beauty of concrete,' he said happily, 'you can never be sure exactly what you're getting. We couldn't do it again if we wanted to.'

The birch-faced ply, he said, had added something he hadn't anticipated, 'because there are very fine little fissures in it and as a result the water absorbency changes over the surface, so that although it looks smooth at first glance, when you inspect it, you see that there's this little flecking all over it: beautiful. This is Ivy Grove Grey. There will never be another concrete like it, because this is a handmade house.'

The second wall was on the other side of the sitting room. Joyce and Ferhan had said repeatedly that they had based this room on Ferhan's elegant, first-floor, three-bay, Georgian drawing room. But now that the two long walls were up, the room was looking distinctly like a corridor. What did they mean by 'based on'? Half the size of? And had they considered, when they did their basing, that it might make a difference where the windows were?

One whole side of this room would be window. Unfortunately, it was the short side, whereas presumably Ferhan's windows were strung, jewel-like, across the long wall of her room, so that light poured into every corner. Ours, I suspected, was going to look more like the light at the end of a tunnel.

Christopher Alexander, Professor of Architecture at Berkeley, has written that 'when plate-glass windows became possible, people thought they would put us directly in touch with nature. In fact, they do the opposite.' For many years, Alexander has been engaged in a project to analyse and catalogue architectural forms in an attempt to demonstrate his theory that the most attractive of these are not a subjective matter. They are a delight for all time, because they share certain essential attributes with forms in nature. His 1977 book, *A Pattern Language*, attempts to lay down rules that he derives from these principles. Some of these seem rather obvious – people prefer houses in which there is plenty of natural light – and some are beguiling because they seem to make sense, although you probably wouldn't have thought of them yourself (as a non-architect, anyway): 'The life of a public square forms naturally around its edge. If the edge fails, then the space never becomes lively.' He claims that everybody likes window seats, and that rooms with windows on two sides are cheerier, both of which may well be true. But much of the book is a long howl of protest against modernism. And we were doing nearly everything wrong. 'A building in which the ceiling heights are all the same is virtually incapable of making people feel comfortable,' he writes severely. 'Houses with smooth hard walls made of prefabricated panels, concrete, gypsum, steel, aluminium, or glass always stay impersonal or dead.' I could only hope that Professor Alexander's Casaubon-like project was misconceived.

All the cages for the reinforcing went up in the next few weeks, so that you could soon see where the downstairs walls were going to be. They were, indeed, all going to be concrete. And, as I'd feared, there didn't seem to be enough space for them (and presumably they'd be even bigger once they were sodding great walls and not just delicate meshes of steel). Freddie said whenever someone wanted to come into the kitchen, someone else would have to leave.

I thought about remonstrating with him. I considered protesting that this house wasn't one of those things that adults did – like having relationships and careers – that somehow *relegated* him; but I looked at the cages and my confidence ebbed away. Instead I took his £20 bet that he'd be able to cross his room in fewer than ten paces and privately tried it out (on the floor below, obviously: we were still a long way from getting upstairs) only to find that you could cover the distance comfortably in fewer than three.

Everyone seemed to be anxious. The more the house looked like a house, the less it looked like a house we could live in. Early in June, by which time we had most of the walls downstairs, Harry asked: 'Are we having wallpaper in our new house?'

'No,' I said, 'we're leaving the walls bare.'

'So we're painting them?'

'No . . . er, we're just leaving them.'

'What,' he said incredulously, 'so you can see the concrete?'

'Uh-huh.'

'Ugh!'

There were still design issues to be resolved. Joyce and Ferhan had struggled to fit in the American fridge, wrestling with every conceivable kitchen configuration until eventually they'd managed to find the necessary 90 centimetres. And then I went to the fridge shop and discovered that there is no two-door ice-dispensing American fridge narrower than 92 centimetres.

We tried reorganizing again: we could only get the fridge in if we did away with some cupboards. Then there was another problem: American fridges were not only remarkably wide but deep; it would stick out. I went back to the fridge shop and grumpily chose the most expensive fridge I could find that didn't have two doors or sticking-out bits. It was made by Gaggenau and had a beautiful frosted-glass front with aluminium surround. I asked Joyce to look it up in the catalogue, thinking she would applaud me on my spirit of compromise. She called back to say she didn't like it. It introduced, she said, another material to the kitchen. I was crestfallen – after all these months of Joyce and Ferhan education, I couldn't even choose a tasteful fridge. But then, in the night, I woke up feeling furious. Joyce *was* an

American: how could she not know that American fridges were 92 centimetres? It seemed to me that she thought a fridge is a fridge is a fridge, and the most important thing about it was that it was hidden behind a slab of wood and didn't spoil the line of her units.

A couple of weeks later, I admitted defeat. I chose an integrated fridge. To get the size I needed, I had to pick one with a very small freezer compartment, which meant having a second, free-standing freezer under the counter in the store room. Which, of course, meant even less room for storing things.

The other matter on which I'd had some opinions of my own was lighting. Having dismissed Joyce and Ferhan's suggestion of the pendant light that looked like an artichoke, I had to come up with something to replace it. Sure enough, I found what I considered to be a very fine horizontal disc of opaque glass held in an arc of stainless steel. Elaine and Clive had one rather like it, so I knew it would look good. All the same, I felt rather nervous about presenting it to Joyce and Ferhan because we were not in the habit of making design suggestions to them. And unlike the fridge, I couldn't pretend it was really about practicality.

They said it was overdesigned.

There clearly wasn't really any room for discussion. Chastened, I quickly flicked through the book pretending there were other light fittings in which I was equally interested. I liked a simple amber globe. So did they. They asked to borrow the catalogue. At the next meeting, they announced that it was too big.

Too big? For our elegant room based on Ferhan's? I had never dismissed a light fitting or a lamp before on the grounds of its being over-large. It had never occurred to me in the area of lights that size matters. That, apparently, only started bothering you when you hired architects and let them design you a corridor.

A fireplace, like the American fridge, had been in my original brief but had somehow been overlooked, possibly in the hope that I might simply forget about it. But I kept on asking where it was going to be. By this stage, Ferhan wasn't sitting in on the design meetings, on account of the detailing being so detailed; but to explain the fireplace, she sat down next to Joyce, poured herself a coffee and said: 'So. Have you told them that if they have a fireplace they can't have a television?'

It was a joke, but it was a Ferhan-joke, i.e. we could have a television at one end of the slate bench and a fireplace scooped out of the other if we *absolutely* insisted, but she couldn't be expected to approve of it. Still, on the whole, the design meetings were, as Charlie said, two parts therapy to one part practicality: a relief from the anxiety of visiting the site and worrying about why nothing much had happened, or lying awake wondering where we were going to get the next slug of money from, since the builders' bills were coming in to a schedule unknown to the Woolwich Building Society.

Charlie had managed, just about, to stick to his tax bill schedule, if only by sometimes using money that was supposed to pay the builders. He'd developed a new strand to his career, as a speaker, which helped to top up the money he was getting for his various projects. But our bank account, in those months, must have looked very odd: lump sums coming in and almost immediately going out, lurching from what looked like wealth to extreme insolvency. On the whole though, things were easier now that we had access to money, even if we had to juggle income to pay bills for which it wasn't earmarked and then hope vaguely that we could cope later. The process, as Charlie said, was like being on a roller-coaster: we were so focused on crawling up the next ramp or edging round the next bend that we failed to notice how high off the ground we were.

'Oh, you're going to have so much space!' Joyce and Ferhan would cry intermittently in our therapeutic design meetings, while we, lulled back into the delightful, design-phase sense of optimism and possibility, would nod dumbly and almost, for a moment, believe them.

We started looking at furniture, dragging Harry and Ned through furniture shops in a way that reminded me unpleasantly of the Bakers Arms Carpet Centre. Both boys seemed to think the furniture in these shops was for climbing over and sitting on, when it was obviously only meant to be looked at and perhaps occasionally lightly indented by glamorous adults who were too busy taking business trips to Milan to hang around long enough to spoil the cream covers. None of the furniture looked like it was meant to share house room with people who usually had chocolate round their mouths and honey on their hands. Still, we did eventually find a sofa with deep sides for the den, on which we could imagine three or four of us snuggled down watching television early on gloomy autumn Saturday evenings. We measured up. It was way too big for the room. And it wasn't even as if this particular sofa was especially large. In the shop it had looked quite modest. We checked, and rechecked. We measured, and did sums. We wondered if perhaps we were in millimetres when we should have been in centimetres. Apparently not. We put it aside for the time being and found another sofa, for the sitting room this time: L-shaped, which we thought might mitigate the impression of a long thin waiting room. The long part, inevitably, was absolutely fine, but the chaise sticking out of the bottom would have hit the wall. (This was the sofa that Joyce said was overstuffed, so we couldn't have had it anyway.) I lay awake at night wondering how a brief for a bigger house could possibly have issued in a house that was too small to furnish. Charlie said he supposed it was rather late in the day to have got out our rulers.

The ground-floor walls were poured at the rate of around one every week to ten days, which was a lot slower than on Ramesh's latest, ever-optimistic schedule. Still, the pace was slightly less painful now that there was something to show for it. We were ready to pour the first-floor slab by early June; according to Brian Eckersley, this was probably the most technically troublesome aspect of the building, because it had to span 7 metres of kitchen, between two walls that didn't even reach its outer edge, then cantilever out to the garden. Since there were to be no columns or beams, all the support for it had to be contained within the structure itself. As a cementitious material, Brian explained, concrete is very strong in compression but weak in tension; in other words, it will take heavy vertical loads on top of it, but string it across a big distance between two walls and it has a tendency to sag. That could be solved only by the organization of the reinforcement: it made a difference whether the bars were at the top or bottom, pointing towards the garden or the lane, and how they intersected; and that was both a complicated mathematical calculation and a nightmare to draw.

But the slab went up, supported, initially, by vertical poles. And then, just as the house was beginning to look slightly more like a house, less like a druidic arrangement of megaliths, things mysteriously ground to a halt.

Joyce went away, initially to celebrate Ferhan's birthday on a boat on the Bosphorous and then for a week on a Turkish beach; by the time she came back in early July we had drifted past our original deadline for the house to be completed and still had no upstairs walls. It was difficult to see any progress while she'd been away. When she called Ramesh she had to ask him to shut up about her holiday because she didn't want to be pleasant.

It was agreed that there should be a sort of crisis summit meeting at the site on Tuesday, 9 July. By then, Varbud had managed to pour one of the upstairs walls. But that still left all the rest. Ramesh opened the meeting by muttering gravely about how difficult this project had been, how much he'd underestimated the time it would take to build the formwork, and how much money they'd already lost. (Joyce and Ferhan also claimed that they'd lost money and, unless Joyce's time was worthless, I couldn't see how it could possibly be otherwise.)

Charlie and I couldn't get Varbud straight. We couldn't work out whether they were dreamers, craftsmen pottering away at their own pace in a mist of goodwill, the sort of romantics who might move halfway across the world to open a drive-in cinema, or whether they were clever and calculating and just very good at making this look like an Indian drive-in thing while they played us along.

Ramesh laid out a new plan of action: they'd pour the second upstairs wall tomorrow, another the following Tuesday or Wednesday, and then bring in a new team to start building the reinforcing for the roof slab at one end of the house while they continued to build the formwork for the vertical walls at the other end. In this way, he expected to have poured the roof slab by 26 July. And after that, it would take only six weeks: his people were used to fitting out thirty flats in that time, so one (small) house shouldn't present a problem. In any case, he said, we shouldn't worry about the recent lack of progress, because they were doing a lot on the joinery back in the workshop.

Charlie said something about the importance of getting it right. (As Joyce, Charlie and I all kept neurotically reassuring one another, Varbud were taking a great deal of care. And we were conscious that if we applied too much pressure we could end up with shoddier work.) But he also explained about having sold our house to the people who were hanging on for us, and that having created some urgency.

Still, if Ramesh's predictions were correct, Varbud would be finished in early September. A few weeks earlier – before the recent vegetative period set in – Joyce had predicted that they wouldn't be finished before late September or early October. But even if we added a month to Ramesh's schedule, that was still OK, because there was all the chaos of the summer holidays to get through between now and then. Soon the house would be full all day every day, and, between the demands for money and attention, and telling people to turn their music down and stop fighting, I'd be finding it difficult enough to work, without having to move as well.

I asked Joyce quietly what she thought of Ramesh's timings, and she shrugged.

No one mentioned the penalty clause. This was partly because we'd now accepted so many different schedules that we weren't entirely sure when it was supposed to come into operation. But we were also so far from being finished that if we were to impose it now, I feared we could make Varbud bankrupt. And then the house would never be finished and we'd go bankrupt too.

Meanwhile, in the time that Charlie and I had been focusing on the house and apparently making very little progress, Hen had been having a whole lot of new and vivid experiences, not all of them pleasant (in fact, most of them quite disillusioning), including acquiring a fungus that it would take her years and a stringent diet to get rid of. She finally arrived back at Heathrow at dawn one Saturday morning in June: not terribly well, and dreadfully saddened by the death back in England of one of her friends – which was why she was back earlier than expected – but relieved to be home.

When she'd left, I'd thought we might be in the new house, or at least on the point of moving in by the time she returned. It was now clear we wouldn't be in the new house until after she'd started at university. But even then, I told myself, she'd come back for long stretches; I just had to hope that my anxieties about size were exaggerated. The last thing I wanted was for her to get the impression that the house wasn't big enough to accommodate her, or somehow discounted her.

Elaine's house, irritatingly, *was* finished. It was also huge. She not only had a World Cup Viewing Room but also enough cupboards in which to store things like tents and fishing nets and table football that modern families seem to acquire, but which we, in future, wouldn't be allowed. Every time my family went out – there was a very useful music class on Saturday mornings – I surreptitiously threw away things I thought they wouldn't miss.

The other irritating thing about Elaine's house was its pristine quality. I would go over and look at her Poggenpohl kitchen with self-closing drawers, at her bedroom that stretched the length of one wing of the house, with Christopher Alexander-approved windows on two sides, and I'd have to fight down my feelings of envy. Then I'd go back across the road and stare vengefully at our stained sofa, where people had dripped yogurt on the arms, and our garden that was a mess because for three years we'd been thinking about moving and no one had paid it any attention.

I was reminded of my mother, who had also, for some years, lived opposite her best friend, my 'Auntie Grace', who had had painted nails and didn't really believe in housework. When she'd lived in Leicester for a time, she'd recklessly driven down the M1 every week to have her hair done at her usual place in Woodford. Yet despite this diva-ish demeanour, her house was always immaculate. My mum would try to console herself with the thought that if you looked carefully, there was dust on the ornaments, but it didn't really help. Auntie Grace had a tidy husband and son, and my mum had an untidy husband and two messy daughters and was not a diva. She never stopped doing housework, but it was a war of attrition in which the mess was always perilously close to winning.

As a child, I couldn't understand why the pristine quality of Auntie Grace's interiors bothered my mum so much, because it seemed to me self-evident that our house was preferable. But now, visiting Elaine's, I experienced what I think was probably an identical sense of personal failure, a deep disappointment that my domestic management looked, by comparison, so inept.

One of the things that had been most satisfying about the

house project up to this point was its joint nature. It wasn't my little hobby or Charlie's pet project. Betty Friedan identified the house as a domestic trap, but for my generation of women, with rich lives of jobs and children and partners who want to be fathers and are not ashamed to cook, it doesn't have to feel like that. Houses now seem much more like androgynous meeting places for people in relationships of equality. In 1879, Joseph van Falke wrote in *Art in the House*:

> The husband's occupations necessitate his absence from the house, and call him far away from it. During the day his mind is absorbed in many good and useful ways, in making and acquiring money, for instance; and even after the hours of business have passed, they occupy his thoughts. When he returns home tired of work and in need of recreation, he longs for quiet enjoyment and takes pleasure in the home which his wife has made comfortable and attractive. She is the mistress of the house in which she rules, and which she orders like a queen. Should it not be then specially her business to add beauty to the order which she has created?

Home used to be the place men went out from, to find out who they really were, to be manly. You couldn't be Odysseus or Jack Kerouac in the kitchen. Now Charlie works from home and regards the house as his preoccupation and project as much as mine. But even though there's no longer anything wrong with men being domestic, with not merely caring about their surroundings but also taking responsibility for them, I suspect a sense of Domestic Standards still hangs more heavily over women. We feel more acutely any lapse; we unreconstructedly assume anything squalid or disorganized to be fundamentally our fault. Charlie didn't think he was a failure because there were piles of newspapers behind the sofa or unsorted bills and letters from school piled up in the corner of the kitchen. He might have felt vaguely irritated by the kitchen drawer that kept sticking, but he didn't think it was supplying some sort of moral commentary on him. The notion that the house is somehow an expression of the self can be rather burdensome for working women, because there simply isn't time to do everything properly.

And my house was definitely out of control. I could make the excuse that I'd neglected it to focus on the new one. But I didn't see how things could possibly be much better after we moved – not, anyway, unless I could somehow get rid of approximately 90 per cent of my family's possessions.

10

Once we were well into the building process and it was too late, Joyce and Ferhan admitted that, months earlier, they had been approached by *Grand Designs* and had turned them down on the grounds that people pointing cameras would have been 'too much of a distraction'.

This was obviously very disappointing: I could have been having all those conversations about the size of the slug garden with Kevin in person, in his own words. But it did give me an idea. I formed a Kevin Backup Plan.

Grand Designs was about to start its third series. I decided to interview Kevin, and I decided to do it at the site. It would be like my very own edition of *Grand Designs*, without producers butting in ('I think that's quite enough on the size of Freddie's bedroom now, darling') or girls wanting to powder Kevin's nose just as the discussion of modernism was getting interesting.

Kevin was agreeable, so we set a date. Unfortunately, I had omitted to plan for rain. Since one whole side of the house was exposed, and it wasn't a large building, this meant that being inside was really like being outside. Also, there was nowhere for us to sit. And even if there had been, we'd have been in the way. Added to which I almost certainly wouldn't be able to hear Kevin's voice in the tape recorder over the noise of the Asian radio station.

Only after I'd arrived at the site did I realize that, given the drizzle and these attendant difficulties, we'd have to do the interview in the car. I hoped Kevin wasn't too starry: plenty of television presenters might object to being dragged over to see you, rather than the normal thing of you politely coming to them, especially if they were then interviewed in a car.

I arrived first. The lane and the site, now that I came to look at them through Kevin's eyes, seemed rather unprepossessing. There were a lot of strips of blue plastic tape snaking across the ground. The whole area was strewn with lumps of concrete, blocks of wood, bundles of wire, empty cement bags and rubbish: an empty Lucozade bottle, a trail of KP Nuts wrappers and a bewildering quantity of peach-coloured tissue.

Still, I was excited by the prospect of Kevin's reactions. I was hoping he would reassure me that such daring walls and radical design made running out of money seem unimportant.

Kevin duly tramped down the lane, only a little bit late, as tall, handsome and muscularly articulate in the flesh as he appeared on television. We shook hands, then I said self-consciously: 'Right, so, er, this is my house,' leading him on to the slab, aware all of a sudden that it didn't look like a house at all, but one of those half-finished buildings you see in Croatia or Turkey with chickens running around in the rubble.

'Hmm,' he said.

'And this is the concrete,' I added unnecessarily.

'Why concrete? Why not timber? Why not brick?'

I began to explain the history of my relationship with the concrete – how the architects had wanted it initially, how I'd been afraid that it would look ugly and brutalist and multi-storey car park . . .

'But presumably you're putting internal finishes on it?' Kevin said, in much the tone of voice that Harry had asked about wallpaper.

'Well, um . . .'

But he'd moved off. My practised little spiel about how tactile and sensual I found the concrete seemed to have been undermined somewhat by his evident impression that the idea of living with raw concrete was deranged.

'I'm worried it's too small,' I told him instead.

He said consolingly that buildings appear to be different sizes at different stages of their construction. 'It'll look bigger once it's painted.'

'But it's not being fucking painted!' I wanted to shout. 'It's a concrete house! The whole point of it is to look concrete-coloured!'

'Hmm,' Kevin said again, having covered the length and breadth of the slab in a couple of minutes. 'I'm afraid I can't comment on its "contribution to the built environment".'

Why was he saying this in inverted commas? Did he think the house was funny? What precisely was there to be ironic about? 'Perhaps,' I suggested feebly, 'since it's, er, a bit wet in here and stuff we should go and do the interview in the car?'

There were, after all, questions I wanted answered, issues that I believed that Kevin might be able to resolve. Why, for example, did everybody ignore the moral of *Grand Designs*: that a project always overruns and the money always runs out? Why, when I told people I was building a house, did they invariably say, 'Oh, how fantastic!' despite the fact that it wasn't fantastic at all?

Kevin said that was because people saw building a house as an adventure. But it was an accessible adventure, one you could have without giving up your job and taking the kids out of school, unlike, say, driving a minibus across America or paddling a canoe across the Pacific (i.e., it was an adventure for people who aren't very adventurous).

'The worst people,' he said severely, as if I'd asked, 'are those who want to improve the built environment, to leave something behind.'

I experienced a brief moment of panic: did he think this was me? Was that the reason for the inverted commas? It had, of course, crossed my mind that the house would probably survive me. Buildings, it has been said, are a way of conducting a conversation across the generations, and that seemed to me a nice conceit. But I surely wasn't having some overblown Ozymandias moment? I just wanted to live in a house where Charlie didn't have to work in our bedroom.

'The other worst ones are the anoraks,' Kevin continued severely. 'People with obsessions – for building underground, or promoting some agenda of their own.'

I definitely didn't think I was one of those. I was painfully short of agendas. I did what I was told. I was, arguably, mildly obsessed with concrete, but in a harmless way.

'I'm amazed that people ever bother to write in,' Kevin added. 'You know, they've seen the series, they understand how people's failings are writ large across the screen. You'd think they'd run a mile.'

'But you're very nice to them,' I said hopefully; this was nothing like the conversations we'd had in my head. 'You discuss it all very sympathetically.'

'I do; but I'm changing slightly. I'm becoming a little more open about that. What fascinates me about these projects is the way that human nature is exposed. People write in and say, "We don't believe how appallingly run these builds are that you film. You must have a bit of an agenda going on here, and we're going to put the record straight." It's incredibly foolish: they're revealed as just as inept, just as incapable of holding on to it as everybody else.'

The other thing about self-builders, he went on (blimey, there was no stopping him once he got going), was their maniacal ambition. 'Once you let ambition out of the bag it runs riot. People will have £200,000 and rather than build a £180,000 house and have £20,000 for contingency or furnishings, they spend £210,000 and have nothing to sit on, and have to remortgage and sell their car. It makes me laugh.'

So there it was. Risible. I now knew that I was a person in need of adventure without being very adventurous, plus over-ambitious, plus mildly delusional. 'You couldn't find an event or project or occasion in life which is so replete with optimism. People will delude themselves to the point of bankruptcy. The builders can have left the site and the mortgage company be repossessing and they still think they're going to live there. I'm terrified by what I see, by the power of self-persuasion.' He sighed. 'Self-builders invest an awful lot of emotion and ambition and hope in their projects. Somehow they displace all their hopes for the future.'

What did he mean by that? That it wasn't actually going to be all right for my family in this house? That this was somehow a way of avoiding the real, hard work of knitting us together? Displacement activity, a way of pretending to myself that everything would be for the best in the concrete future? Reluctantly, I was coming round to Kevin's view that Joyce and Ferhan were probably very wise to have turned down the opportunity to feature on *Grand Designs*.

After we'd finished the interview, Kevin had to dash across town to fulfil some other media commitment. Since we were already in the car, it would have seemed churlish not to give him a lift some of the way. 'If that whole wall is going to be glass,' he said thoughtfully as we queued to get down Upper Street, 'you should make sure your contractors have already ordered it. K Glass is currently taking sixteen weeks to come from Germany.'

Were we having K Glass? What was K Glass? I felt, as I told him, rather inadequate next to his *Grand Designs* people, many of whom were building their houses in an altogether more literal sense, spending their weekends moving breezeblocks or designing composting toilets. I said sheepishly: 'My main role in the process seems to be to look at it.'

Kevin was actually very nice about this. 'You shouldn't underestimate the importance of looking at it,' he said. 'Where you live now, maybe it's an older house, total chaos – and here you will have lots of storage, throw things out, be a modern girl, live in the twenty-first century – and that's a major change.'

He could tell! Just by looking at me he could tell I was dyspraxic and lived in a mess and that to move here I'd have to throw away most of my possessions. He could tell that my modernist fantasy was fraudulently unsustainable.

Still, he had liked something. He had lingered in the sitting room over a pile of yellow foamy board stacked high on the floor. 'This is very good insulation,' he said approvingly. 'It slots together.'

One ordinary Thursday morning in August, I carried out an audit of all the items that were currently out of place in our house. I did not include that day's newspapers, toys that were conceivably being played with or anything that was on the floor in the process of being used. This proved that there was an average of eight things in any room that didn't belong there. Non-rooms, such as hallways, seemed particularly prone to acquire rubbish. On the landing outside our bedroom there was a pile of washing, a fishing net, a bag of Harry's new school uniform, a sewing kit and a large plastic laser gun. In the downstairs hall there were three buggies, a skateboard, a stuffed Piglet, a dressing-gown cord and a pink plastic cricket bat. We were keeping a small plastic Goofy and half a Lego Technics model by the kitchen sink, while the kitchen floor was apparently being used to store a roll of Sellotape and one flip-flop of Hen's. The work surfaces were cluttered with toy cars, postcards, and a plastic tray containing tiny pots of modelling paint. I couldn't even remember what the model had been, but hadn't been able to bring myself to throw the paints away because the colours were unusually winsome – ochres and dusky pinks.

There were two ways of looking at this: we had too much stuff, a lot of it rubbish, or I was a slut.

Certainly, I was lousy at throwing things away. In 1890, in an early discussion of the concept of the self, the psychologist–philosopher William James wrote, 'It is clear that between what a man calls me and what he calls mine the line is difficult to draw. We feel and act about certain things that are ours very much as we feel and act about ourselves.' I was incapable of restricting this to certain things. I told myself that Harry, or maybe Ned, would play again with the motorcyclist who had nothing to sit on, and there was bound to be an occasion when I'd be desperate for a thimble-sized pot of taupe paint.

You also had to consider that we had now lived in our current house for ten years. They'd been pretty good years, and they'd resulted in the accumulation of all these possessions. Simply to have chucked them out would have felt like denying the significance of any one of a hundred million moments, all of which I wanted to hang on to. Looking at these Goofys and laser guns, touching them, the past seemed almost graspable. And it wasn't as if the children wanted me to get rid of things. Quite possibly they felt that I had already shown myself too eager to slough off the past.

Charlie said I should leave it to him because he was good at throwing things away. He once threw away some sleeper train tickets home from Italy because they looked untidy on the villa dressing table. So that was probably an even riskier option.

One morning he retrieved a small duck egg-blue pottery vase from the back of a cupboard and suggested getting rid of it (it had originally been his).

'No, I don't think so.'

'But we've never used it.'

'All the same,' I said worriedly, 'we don't have any other vases that size.'

My grandparents, who probably acquired fewer possessions in a lifetime than I did in a year, couldn't afford to throw things away and kept almost everything they'd ever had – old cake tins, place mats, ornamental wooden animals from the 1950s – preferably wrapped in polythene bags and elastic bands and stored at the back of a drawer. The aim of all this keeping was to hand the things on to needier people, i.e. me. Whenever I was offered them, I felt like a rampant consumerist ingrate, even though the truth was that I had quite enough tat of my own without taking on someone else's. But you can't tell someone who has carefully kept a biscuit tin for twenty years so that eventually they can pass it on to you that they shouldn't have bothered. These possessions were rich with memories, and that's what they were trying to pass on, the feeling of existence being shored up.

I wasn't sure that I liked the values of throw-away people anyway. If you could chuck out all this consumer junk so easily, had it been worth having in the first place?

My sister walked around the shell of the new house and said dubiously that we could always use paper pants. What I really needed, it was pretty obvious, was two houses, in one of which sentiment was exalted over aesthetics. I felt mutinous about what was entailed in being what Kevin called a modern girl. So rather than get out my dustbin bag and scoop up the Goofy, I stomped around muttering about the perils of living in a visual culture. Soon, I muttered darkly, I supposed I wouldn't even be able to look tatty myself. No wrinkles allowed; I'd have to get facelifted so I could look all shiny and new to go with the house. There would be no acknowledging how I got here, at what cost, with what damage.

Charlie ordered a skip. Even I could be persuaded that in the leaky shed or the area with the damp problem under the front steps, there might be a few things that could be relinquished. But no sooner had the first of these items landed in the skip than people started landing in it too, sorting through the rubbish like children on rubbish dumps in a favela. These, though, weren't children, or poor.

'There's nothing valuable in there,' I said mildly to a hippyish bloke.

'Not to you, maybe,' he said without looking up, which I found a bit offensive; I think I'd have known if I'd had a Monet stashed under the front steps. Charlie and I had inherited quite a few biscuit tins, but sadly no major works of Impressionism.

It is obviously ecologically preferable to recycle the rubbish to people who actually want threadbare blankets and ancient stained pillows than it is to stick it in a hole in the ground. But I still felt unaccountably irritated. Charlie said he didn't mind them taking things; he was more offended by what they left behind. Why, for example, had no one leapt on his book about the postwar history of the Japanese economy?

It was the hottest day of the year, with a heat haze rising off the tarmac, Hackney listless with discomfort, our rubbish glinting malignly. Charlie and I worked all day and filled up the skip, probably twice, considering how much got taken out. (What did the hippyish bloke do with the black plastic flowerpots? Did the rusty tins ever come in handy? He should have met my grandmother.) By the evening we were exhausted. I looked around the house. We seemed to have made no impression at all.

Just occasionally, now and then, when I visited the site, I felt the same little spurt of excitement that I used to have in design meetings. It was admittedly regrettable that the garage was the biggest room in the house, but at least the underfloor heating was in, a snake of tubing doubled back on itself concertina-like across the floor. Heating, unlike concrete, was a thing that real houses had. Every now and then, just occasionally, I believed I might live here; I had the slightly breathtaking sense that this was the house we'd designed.

At the end of August, Varbud invoiced us for £25,000 and warned that they'd need a further £50,000 in a fortnight. We didn't have any money to pay these bills; we hadn't even had any money to pay a £60 bill for dinner the week before. (We'd offered up three dud cards before managing to produce one that was creditworthy, while the French waiter's 'Je suis desolé' became progressively more ironic.)

We knew that we would be able to draw down the next slug of the mortgage as soon as the glazing was completed. But even assuming that the glass was ordered, which I suspected it wasn't, and that it wasn't going to take sixteen weeks to arrive from Germany, which I suspected it was, Varbud wouldn't be ready to fit it for weeks. We could draw down a further £90,000 beyond that, once everything was plastered, even though there was hardly any plastering to do.

Charlie asked Joyce if there was any chance that she might sign off the next stage of work regardless of the fact that we were nowhere near having done it. Not unreasonably, she was affronted at being asked to lie. It would be utterly unprofessional, she said; and what would happen if Varbud went bankrupt?

The trouble was, unless we got hold of some money soon, we'd all go bankrupt. Charlie didn't like having to ask Joyce to lie, he didn't like her pointing out that it was wrong, which he knew, and he didn't like her refusing. So one way and another they had a very sticky conversation which ended with Joyce noting acidly that this (our not being able to pay) seemed to be happening more and more often. Which we also knew. Afterwards Charlie said he didn't want to have another meeting with her for a while.

He tried Steve, who was now back from his mystery illness. Charlie was all borrowed out, but Steve thought he might be able to raise a loan on my account. He said gloomily that he'd 'have to see what security I can offer my risk people'.

Charlie also asked for a meeting with Ramesh, which, luckily, turned into a sort of love-in. He came home saying dreamily that they'd agreed that they must look after each other. (I'd long thought the builders must be on drugs; maybe Ramesh had slipped him something.) Ramesh would accept payment in stages if necessary (if I did get a loan, it wouldn't be for much), while Charlie promised that they would have a bonus – though I don't know where we were supposed to get this from – if they could get us in by 7 December.

Ramesh said he thought they'd have done the plastering in the next fortnight. So we agreed that we'd try to persuade the Woolwich to swap the payment stages, so we could have the plastering money first, then the glazing. (Joyce, who was still grumpy, said she doubted the Woolwich would release money for plastering before the building was dry, and if they wanted her to say it was dry, she still wouldn't be able to sign it off.)

House building, Charlie said at the end of this meeting, was a trade in lies. Ramesh said the glazing would be done by the end of September. (He insisted the glass was only taking three weeks to come from Germany, not sixteen, and that it took only a day to fit. You just needed to order the crane, he said breezily. I trusted he'd factored in a few days for forgetting it.) Charlie said we had the money to pay them. And Joyce said we'd have enough space.

As Charlie pointed out miserably one morning after yet another sleepless night, building a house also made all the things that ought to be unimportant, like bricks and mortar and money, loom in the foreground. And all the things that really mattered, that should have been the substance of our lives – work, experience, the quality of relationships – got forced into the background. It was relentless and grinding, and at the end of it all, we couldn't even have the fridge we wanted.

We should, we agreed gloomily, have moved into a big old cheap house, done it up and gone on a lot of expensive holidays.

11

The people who'd wanted to buy our house the previous Christmas were still hanging on in there, despite the alarming elasticity of our plans. Believing our assurances that we would be ready by summer (which to be fair, we did too) they had put their house on the market in the spring. And, as is the way of housing chains when headed by someone who flatly refuses to move, by August both they and their buyers were in danger of having the deal fall through.

Someone further along the chain volunteered to rent so that we could all link up eventually. In return, they wanted a contribution from everyone else along the line to their costs – £2,000 appeared to be the going rate – and a fixed completion date of 5 December.

We went away for a week while these negotiations were taking place, during which the tense discussions on the mobile phone at least made a change from being refused money at the ATM machine. We managed to push the completion back to the 19th, although when I called Joyce to tell her (i.e. to make bloody sure Varbud were finished in time), she said unencouragingly, 'Well, I certainly would hope . . .'

Two days later a general purchasing revolt got the date pushed forward again to 12 December.

As soon as I got home, I headed for the site to check on progress.

'We're about to do the scree,' Mavji announced.

'So then you'll be able to get the limestone down?'

'Well, yes, once it dries.'

'Right. How long does that take?'

'Five weeks.'

According to Ramesh's schedule, the limestone was supposed to go down two weeks after the scree. Which made me wonder how much of the rest of it was fictional. Mavji assured me there were other things they could be doing in the meantime, such as joinery, in the workshop.

Soon after we got home, Tom Tasou sent us a letter pointing out that the extended deadline for reinstating his electronic gates had expired in February 2002, which, we might like to note, was now six months ago. Accordingly, he had decided that if the work was not completed to his satisfaction within twenty-one days, he would instruct the builders to put the gates back in their original position, revoke our agreement, and send us the bill.

The main reason why the gates hadn't gone back up was that Varbud still had to pour the front wall of the house – actually the outer skin of a double wall, with insulation in the middle. This was a technically difficult operation a) because the wall was only 150 millimetres thick (not thick at all), and b) because what with the inner wall and insulation, you could only get at it from one side. Varbud had also decided not to use diagonal buttresses, which would have blocked the lane. It was bad enough for the tenants in The Glassworks not to have any gates; they'd have been even less enthusiastic about not being able to get past at all.

When Varbud did get round to pouring this wall, the formwork wasn't secure enough (no diagonal buttresses) and the concrete spilled out of the bottom. My brother-in-law Clive, who had gone down the lane for a look on Saturday afternoon, reported that there was an enormous amount of activity on site, which I couldn't understand (Varbud never worked Saturday afternoons) until I realized they'd had to knock the wall down and start again. VM said next time he'd put up some diagonal buttresses 'just for a few hours'.

Meanwhile, Joyce and Ferhan called a meeting to warn us we'd already spent all the contingency. This had gone as follows.

£3,500 on extra formwork (Varbud admitted they'd grossly underestimated this, and claimed they'd shouldered £10,000 of the overspend themselves).

£500 on the hoarding to protect the roots of the ash tree (made of several sheets of chipboard, which looked like it cost, at a generous guess, £60).

£2,000 on additional limestone (where?).

£450 for a window in Hen's room, which Joyce and Ferhan claimed had been required by the Planning Committee (except that we'd got planning permission in April 2001 and Varbud's estimates on 25 July 2001, which had allowed them roughly three months to work out that it was there. The window was useless anyway, since it was too high to see out of.)

The rest was on provisional items that cost more than anticipated (e.g. an extra £450 for the electronically operated rooflight over the stairs) and a few things that we'd decided to upgrade, such as towel rails.

What it meant, though, was that we had to make some decisions. The shutters, which were supposed to go on all the upper windows, had turned out to be more of a feat of engineering than anyone had anticipated. Estimated at £6,700, they were actually going to cost £14,000.

The fireplace had never been costed and had come in now, very expensively, at £3,000 (at this point I kicked Charlie under the table).

Then there was the roof. We could, Joyce explained, have an asphalt roof at the price quoted, or a state-of-the-art roof, guaranteed for ten years.

Actually, ten years didn't sound that long to me. And if that was the better roof, what would happen with the other one?

'Varbud would come back and fix it, obviously,' Joyce said airily.

At home, I fished out my copy of the spec, a fat, largely impenetrable document dense with letters and numbers, product codes and purchasing references – clotted details that if you got them wrong, could leave you with no windows.

Joyce appeared to have specified something advertised as 'the ultimate flat roof', made of a material that sounded like a character from Star Trek, called Evalon C. I rang her and said that was good enough for us.

It turned out that the ten-year roof and Evalon C were one and the same. Varbud had allegedly been told that it could be had for the same price as asphalt.

'Then it'll have to be asphalt,' I said grumpily.

A few days later, Joyce announced that an asphalt roof would require some kind of parapet around the top of the house, which would look funny. And she'd spoken to the Asphalt Council and it would have to be painted silver or white – something to do with heat and reflecting the light: maybe we should at least have the super-roof on the garage, since we'd be looking down on it? Oh, and she'd found £1,500 from insulation that we wouldn't need if we had the expensive roof, because insulation was already included in the package.

We gave in. I was aware that a major factor in not choosing the expensive roof had been that it wouldn't be visible, and this was a bad reason. I also knew roofs were inherently risky. (When the owner of Fallingwater complained to Frank Lloyd Wright that the roof leaked, he is supposed to have said, 'That's how you know it's a roof.')

Anti-modernist tracts made much of the absurdity of the flat roof, which they saw as the triumph of mad theory over common sense. (Tom Wolfe, in his brilliant polemic, *From Bauhaus to Our House*, notes that the Hartford Civic Center Coliseum had a flat roof in defiance of prevailing weather conditions; as soon as the snow came, this roof 'collapsed piously, paying homage on the way down to the dictum that pitched roofs were bourgeois'.)

Flat roofs, I had read in these books, are easily damaged by people walking on them, by flashings not properly installed; by erosion, sun oxidization, blisters from expanding moisture in the roof layers, splits, ridges, punctures and fish mouths (where the seams between the felts open up). When water comes in through a flat roof, you can't always tell where it's getting in, because water can travel considerable distances in the roof layers. And if you put a new membrane on top, it can weigh down the structure and make it concave. Which leads to ponding, which makes leaks inevitable and untreatable. I didn't want to have to think about the roof. Not for ten years, anyway.

So we capitulated over the roof. And then we capitulated over the limestone. The extra was for outside the house, Joyce explained, although we were already having £3,000 worth of limestone outside, and I couldn't see how we could possibly need two-thirds as much again. She said that she and Ferhan had enlarged the terrace round the house – although we hadn't seen any new drawings – and Varbud had mistakenly assumed the path through the slug garden would be concrete slabs, not limestone paving. (In fact, when the garden was finally designed, the path became gravel, and we had to pay for it all over again.)

Joyce also continued to lobby over what she called the architectural importance of the shutters. Charlie met her to discuss the hanging of Tom Tasou's gates while I was doing an interview in Berlin. Brian Eckersley, who joined them, expressed his amazement at how much we'd got for our money; he was astonished that Varbud hadn't just said halfway through that concrete was too expensive and we'd have to have blockwork.

Fundamentally, Charlie and I thought this, too. But from where I was sitting that afternoon, on a hard seat in a departures lounge with a streaming cold, knowing I wouldn't be home for hours, the architectural importance of the shutters was just irritating.

Joyce and Ferhan presented us with another of their difficult choices over the upstairs floor. They showed us an ugly veneer floor that fitted our budget (we had to have veneer rather than solid flooring, because of the underfloor heating), then a beautiful elm one at twice the price.

Which is how, a couple of weeks later, I came to be up the Holloway Road meeting Robin Hodges, who was to make our floor and who wanted to make sure we realized what we were getting. Elm is very un-uniform, and he'd had clients complain in the past that it looked patchy.

As Robin whisked me around the veneering process, explaining how fussy he was about moisture content and about fixing without nails or screws, he also told me he only uses fallen parkland timber (apart from oak, which comes from the biggest oak forest in Europe, first planted in Napoleonic times, and is cut at the rate of one-third of 1 per cent a year). All his elm comes from Scotland, although there are concerns that with global warming the beetle that carries Dutch Elm disease may move north. He likes the elm best, because it shows how it's grown: you might get greenish mineral streaks in the grain. He refuses to use any tropical hardwoods.

'So,' I said carefully, 'does that mean you can't ever be sure any tropical hardwood's OK?'

'It's all right if it's plantation timber, and they're planting two or three trees for every one they take down. But plantation timber looks the same as any other when it's cut . . .'

'Then you can't?'

'It's as corrupt as hell. You'd have to know the bloke, and the wood would have to arrive with his stamp on it, and even then . . .' He shrugged.

I drove home telling myself that it had been weeks before I could even remember the name of the iroko, and months before I knew how it was spelt. But these were lame excuses. I was a member of Friends of the Earth. Their magazines came through the letterbox from time to time and they had probably had articles about this – 'iroko, the new teak' – if I'd bothered to read them.

Not long afterwards, we heard that the new European Parliament's building in Strasbourg had had its otherwise impeccable green credentials demolished by a strip of iroko that somebody had decided to incorporate at the last minute.

Meanwhile, what with the roof, the limestone and the shutters (they were architecturally important), Joyce and Ferhan informed us that we'd overspent our budget by £54,082.20, and they'd have to charge us 14 per cent of this, as per their contract.

I was still trying to throw things away, telling myself now that disposability was a good thing because a world in which you had to hang on to things for several generations before they acquired any value would be a depressingly static place to live. Looked at this way, it was practically an act of radicalism to throw things away; an embracing of social mobility, of identity that was not determined by what you inherited or could pass on, but was yours to make. I busied myself subversively with black bags.

It was an exhausting business though, and I began to wonder if I was going slightly mad. I lost the Cash's name tapes that my mother, since I was so distracted, had ordered for Harry's new school uniform. Probably they went into one of the black bags. I spent a whole week trying to remember the name of Chrissie Hynde and The Pretenders (for some reason, I'd started to tell Freddie about them but then I'd got stuck because I couldn't remember who they were) even though I have interviewed Chrissie Hynde and even been makeup shopping with her. Charlie said kindly that I was suffering from cognitive overload.

He wasn't much better. He complained that he never had time to think because he was so busy earning money to pay for the house. But since thinking was what he did for a living, that was not a good thing in the long term. We had a row about it. Afterwards he said I gave him courage to do things, which was a nice way of saying that without me he would have been quite well off and not lying awake at nights worrying.

During the row he said bad-temperedly that he thought that from now on we should ban all conversation about cooker hoods. The trouble was, there were so many cooker hood-type worries that they sort of dribbled out regardless, a Tourette's Syndrome of domestic appliances. We could pick up conversations about white goods in mid-sentence. I began to wonder seriously if we *had* any other conversation.

He was right, though: it couldn't be healthy to devote so much energy to every tiny detail of our physical surroundings. I longed to get into the house, so that it stopped being a succubus in my brain, had to retreat into houseness and *shut up*. It wasn't just that Charlie and I didn't talk about anything else: neither did anyone else. 'How's the house going?' people would invariably ask. I had forgotten what people discussed over dinner or passed the time of day with in the street when they weren't building houses. They must have had whole other lives of movies and books and gossip and what was going on in the government. We didn't; and when the house was finished I couldn't conceive of us having anything to say.

I couldn't understand what was so important about houses anyway. It was what went on in them that mattered. Some old friends of mine came to visit that summer, Americans who'd never met Charlie, and he told them (possibly slightly defensively: looked at through their eyes, the house we were living in must have seemed a tip) that we could be happy anywhere. Which I hoped and believed we could. The house was just a shell. In which case, why had we put ourselves through this?

Even at this late stage, we were still acquiring consultants. Since the video and DVD were to be located some way from the television, Joyce recommended that we hire someone to tell us what to do with the wiring. So we called in Mark Drax, who was very tall, looked a bit like Kevin's younger brother, and spoke like a character from an old British war movie. Mark talked about 'sending in my chaps', as if we were the bridge at Arnhem, and referred to Joyce as 'a great girl', as if she were fifteen and goofy rather than a serious professional architect who was mostly, these days, seen arguing with Varbud.

Mark offered to design a system for £50 an hour – he thought it would take about four hours – and then we could buy the equipment through him.

'I suppose you're going to tell me you're competitive?' I asked sceptically.

'You can ring up Peter Jones,' he practically shouted, as if advising a forced march, 'and if you can buy it there, I'll refund the difference. I'll be jolly surprised if you can.'

He could probably be so confident because the make of equipment he specified was one that Charlie and I had never heard of, and it wasn't stocked by Peter Jones.

By October, visible things were starting to happen inside the house. The window frames went in downstairs. The iroko cupboards and shelving transformed the concrete, highlighting its coolness and at the same time warming everything up; you could see that (as Joyce and Ferhan had promised) the space wouldn't feel cold or forbidding. The rosy wood seemed to glow alongside the grey walls, accentuating their smoothness and implacability, but also making them more approachable, less domineering.

The first of the stairs were poured, too, in stair-shaped wooden boxes, although when Mavji put them up against the wall they stuck out too far into the hall. There had been, he said, 'something wrong with the calculation', which made me wonder whether any other calculations might also have gone awry. In the den, for instance, which, on account of its minuscule size, we had decided to leave open to the hall in one direction and the kitchen in another. Since a third wall was glass, it now wasn't remotely den-like. We had finally located the world's smallest sofa to go in there, but that would have to be it for furniture. I no longer wanted to call it the den, which also implied to me that it should be full of soccer moms, when I wasn't sure there was even room for me. We thought about 'snug', but it wasn't; and Freddie suggested cotching room, but that was both too much of a mouthful and too effortfully hip in a way that would ultimately be embarrassing. I worried that these difficulties over nomenclature indicated that it didn't really have a function.

Then there were the bedrooms. 'We've got the beginnings of a bathroom,' Charlie called out from up the ladder to the second storey, on one of our now increasingly neurotic weekend inspections; then added, 'Oh my God!' because the bathroom didn't appear to leave room for a bed.

(These weekend inspections were additionally fraught because we had to take Harry and Ned, who were unmoved by our enthusiasm for underfloor heating tubes. One or other of them would try to impale themselves on something spiky or fall from a great height while we were distracted by room size.)

Shortly afterwards, Joyce and Ferhan announced that they'd decided to turn our bed around. They insisted that their work on the bathroom had given them a chance to rethink the whole master bedroom space, and seemed quite offended when I suggested that there might be an ulterior motive. I remained convinced that it was because if they'd left the bed where it was, we wouldn't have got round the bottom of it.

Joyce badgered us to supply bed measurements for the children's rooms so she could design bedside tables. Charlie ordered some beds from a shop called Purves and Purves and gave her the dimensions. A couple of hours later she rang back to say that Harry's would stick out into his doorway and Hen's room would require a smaller desk.

She suggested I measure out 500 millimetres on the dining table and work out if Hen's laptop would fit on it.

I didn't bother. I thought I owed Hen space at least for a couple of pencils as well as a laptop. She was supposed to be doing a degree. We decided to get the beds made.

'It's good they'll fit snugly,' Joyce said.

Yeah, I thought; they'll be like those boxes at railway stations that Japanese men climb into when they're too pissed to go home.

Freddie, who had been relatively quiescent for a few weeks under the misapprehension that the children's bathroom was actually part of his bedroom, was back on form, as that space started filling up with things that were unquestionably bathroom fittings.

'Why are we having an airline toilet?' he asked bad-temperedly (it was stainless steel, with no visible cistern). 'Are you planning space-saving reclining beds and meal trays too?'

Around this time I interviewed a woman who claimed her marriage had broken up because she and her husband converted a barn. This seemed to me entirely understandable. The piece was about stepfamilies, and made me realize that despite the large numbers of stepfamilies in existence, all the imagery of the family – the positive imagery, anyway – is traditional and nuclear. Any representations of blended families that do exist are anguished. Maybe, I thought, we were having all this difficulty fitting comfortably into a house because there was something about us that was against nature? Maybe we were too unwieldy, too hubristic? Stepfamilies have no iconography: what right did we have to think we could make this work?

With a month to go until we were due to move in, we still had no back wall, no roof, and no glass in any of the windows.

I cornered Mavji at the site. 'OK,' I said seriously, 'when do you think it will really be ready?'

Mavji smiled slightly. 'December.'

I pointed out that December was a long month, which started quite soon. 'You'll have to tell me when in December.'

'Middle,' he said vaguely.

I tried to press him, but he just started muttering about the weather.

Charlie met Ramesh, who was in such a good mood after Diwali that he seemed not to mind that a cheque we'd just written him for £15,000 had bounced. (To make matters worse, Steve had called that morning to remind us that our £40,000 overdraft expired the following week. 'The patience of my people is running out,' he said. Trying not to sound too desperate, Charlie asked if he could at least let us hang on to £20,000 of the overdraft, since in addition to the bouncing £15,000, we were expecting a bill from Varbud for £50,000. We thought we might possibly be able to make up the missing £20,000 with a new loan from Lloyds, now that we'd exchanged contracts.) Anyway, Ramesh promised that the glass was coming on Friday, the joinery the following week. The boiler was about to be fired up, which would get the underfloor heating going, so that the elm floor upstairs could come in the week after next. The bathroom tiling would be done in the next fortnight. So everything was on track. Charlie asked if we'd get in by the 12th and Ramesh answered, 'No problem, no problem.'

Not everyone was so optimistic. On 21 November, we had a meeting with Joyce and Ferhan to discuss blinds for the upstairs windows.

'You're not going to get in for the 12th, are you?' Ferhan asked with her customary directness. 'What are you going to do?'

Joyce, who was speaking to Ramesh on a daily basis, was more influenced by his bland good humour. 'He doesn't want you to rent,' she informed me, after I begged her to get the truth out of him.

'I don't want us to rent either, but that's not really the point,' I said. 'Is he going to put us up when we're homeless over Christmas?'

On the 28th, on one of my anxious daily visits, I spoke to Neven, one of the senior builders.

'Maybe you be in this house for Christmas,' Neven said brightly.

'But we have to be in by the 12th!'

Neven looked at me in alarm; clearly, he'd had no idea there'd been any deadline.

'Maybe you get in upstairs,' he offered kindly, 'and downstairs not be finished. Kitchen not ready.'

Why kitchen not ready? What had they been doing all that time they were supposed to have been concentrating on the joinery in the workshop?

I went home and rang Joyce. '*Why* don't the labourers know what the date is?' I yelled. 'How the hell are they supposed to have a sense of urgency if they don't realize that in a couple of weeks we're going to be homeless? How can there be a deadline if no one knows what it is?'

She passed these concerns on to Ramesh, who called us to say, 'We don't tell the men what we're doing.'

I got the impression I was at fault. Nobody really liked us talking to the builders. Ramesh's view was that I wasn't dealing through the proper channels, so what could I expect?

Which was all very well, but Neven had to do the work, and presumably was reasonably well placed to assess how long it would take.

The glass for downstairs had arrived but couldn't be fitted because the window frames weren't ready (though given the time Varbud had devoted to the joinery in the workshop, they should by my calculations have been finished sometime last spring). Upstairs, for reasons that I was too exhausted to explore, there still wasn't any glass – the panes were bigger, or something, and consequently had to come from somewhere else. When I next found Mavji at the site I demanded to know when this upstairs glass was expected.

'At the moment, they are saying [pause] December 10th.' It was like being told the cheque is in the post.

I paced the house furiously (or the downstairs; it was still impossible to get to the bedrooms, on account of an absence of stairs), thinking that it was as if Ramesh lived in a parallel universe, in which telling us what we wanted to hear was better than telling us the truth. I insisted to Joyce that I wanted a day-by-day breakdown of what would happen until we got in. Tom Tasou had asked for this for his gates because he said he was being made to look a fool in front of his tenants. I was being made to look a fool in front of myself.

Ramesh responded that it would be very time-consuming to draw up such a schedule and would cause delays, and it would really be very much more sensible just to get on with the work.

Until four days before we were due to move, Ramesh continued to insist that we would be in on time. It was increasingly obvious he was talking rubbish. I began looking at houses to rent on short lets. This was a dismal business – most people with nice houses to let want to rent them out for a long time – but I eventually found one in Wapping that was the right size, and not uncomfortable, even if it was in the wrong place. (Being close to the river would be romantic, I thought: the cobbled streets and warehouses, with their alleyways down to the Thames, the dank water and winter fog, the sense of history seeping up the old stone stairways from the river. And all this turned out to be the case, although we never really fitted in. Nobody else seemed to have kids; they all had Porsche convertibles instead. Shopping in the local supermarket was a disorienting business of locating at the very back the few things that didn't come in the form of ready-meals-for-one.)

I took the house in Wapping for three weeks, extendable if necessary. Even when he admitted defeat, Ramesh said brightly that we could store all our stuff upstairs in Ivy Grove Lane. I pointed out we had a piano. There was no way we'd get it upstairs: it would have to go into storage. Not for long, though, Ramesh said cheerfully: we'd definitely be in by Christmas.

The packers arrived in Hackney on 10 December, Ned's birthday (we had to sing happy birthday at his cousins' across the road because by then there were no plates or tables). They spread like locusts through the house, although it already felt half dismantled, because I'd given so much away.

I had hoped to sell a few bits and pieces – a sideboard from Heal's, a couple of limed oak tables – and so I invited a valuer to have a look and give us a price. I don't think he was much good, because he accidentally included in a lot of several items an ebonized Arts and Crafts chair and gave it a total price tag of £150. The chair was by Godwin, and worth £800.

I decided not to bother with him, and to give most of the stuff away to a charity that refurbished old furniture, and employed ex-offenders and drug addicts (who managed to bash the walls with every item they took, and still charged us £15 for our bed, which they said was a disgrace and no one would want). And after that, anything I couldn't bring myself to give away, or to throw away on the tenth attempt, like the Persian carpets, I put into storage. I skipped around harassing the packers, who, I was convinced, were sending the broken stuff to the new house, Ned's new birthday presents into storage and none of our clothes to Wapping.

At the end of the day, I stood in the shell of the house and wondered – even though it looked appallingly scruffy – if we were doing the right thing. All the time we'd been in this house, no one close to us had got sick or died. Harry and Ned had been conceived here. Henrietta and Freddie had effectively grown up here. Their best friends lived across the road in one direction, their cousins in another; there was an easy traffic of children through the houses and gardens. These were the things that mattered to them, not whether their bedside tables were aligned with the bed.

It was also a stupid time to be leaving Hackney. My mum and Clive's dad had sat on the steps of Elaine and Clive's first flat saying bewilderedly to each other that they'd spent all their lives trying to get away from this place, so why would their children want to come back here? But it wasn't like that any more. For Hen's generation Hackney was mostly famous for having the highest concentration of artists in Europe; it was a cool place to live.

There was none of the usual house-moving excitement. Normally, even if emotions are mixed, you can look forward to going somewhere new and presumably more suitable. All we had to look forward to was a meeting with Ramesh to discuss 'progress'.

12

We didn't have enough money to move in. We couldn't pay Varbud's last bill, or Captain Drax for his television; we had no money for the broadband and computer setup or the ridiculously overspecified telephone system.

The only thing we could think of was to increase our mortgage to access some cash, which was all very well, but Steve was on his way out of Barclays on a redundancy package and already on gardening leave. He advised us (we could still – just – get hold of him: he had hung on, for a couple of weeks, to his company mobile) to contact Barclays directly – he was sure the money shouldn't be a problem – but to come back to him if we had any difficulties.

The bank turned us down.

Charlie and I paced by the river in Wapping, adrift in yuppieland, frightened. It now seemed perfectly possible that we'd come all this way and we still weren't going to make it. Despite not actually having been able to afford it, of having no idea what we were taking on, we'd built a house. But possibly one that now, at the very end, we wouldn't be allowed to have.

From the outset – overbidding for the land when we had no prospect of laying our hands on the money – the whole thing had only been possible because of Steve: his gambling instinct, or his judgement of us, or whatever it was had carried us through. The whole venture had been an act of faith, not of calculation. And now he was gone, and people had started calculating.

Charlie's mobile rang. It was Steve, from France, where he was helping a mate of his do up a flat. We stood by the river, Charlie leaning against a parapet, me up against him staring into the swirling brown water, while Steve offered to do what he could. But this seemed much less reassuring than at any point in the past. He didn't even work there any more.

We struggled towards Christmas. The house in Wapping was old and interesting, but suffused, as in the manner of devil possession, with a terrible drains smell. I'd noticed this when I'd been shown around, but had dismissed it as something vaguely to do with the estate agent. The smell was almost unbearable in our bathroom, hung like fog in the bedroom, and kept sneaking down the stairs to the kitchen. It took ages to get something done about it: we told the estate agent, who told the landlord, who had to organize a plumber, who came but said he needed a part that was difficult to source, plus it was winter and Christmas and he was exceptionally busy . . .

Working in the house proved impossible: the desk, too heavy to move, was positioned in front of French windows that didn't shut properly. It snowed that Christmas. And before I could even think about piling on Arctic wear and sitting down at my computer, I had to drive through the festive traffic every morning for two hours. Kerri, who looked after Ned, was also looking after another child near our old house; I didn't feel I could ask her to trail all the way over to Wapping to spend the day in the cold and the smell. So I drove through the rush hour and the shopping traffic, worrying about when I was going to get any work done.

On Christmas Eve, I sat down at the rented, too-small kitchen table and sobbed that I was fed up with trying to be perfect. My family stood about eyeing one another nervously, wondering if any of the others had noticed this was what I'd been doing.

Meanwhile, Ramesh had come up with a new completion date – 21 December – but I was too harassed to contemplate any more upheaval before Christmas; too exhausted from sending out change of addresses in time for 12 December, and Christmas cards, and ordering the turkey and wine, and getting the tree and remembering to rescue the decorations from the packers, and generally carrying off the pretence that we could afford Christmas and afterwards move in and pay the bills.

Some days before 21 December it became apparent that the house wouldn't be ready then, or, since Varbud always took two weeks off for Christmas, any time soon. That Boxing Day, we had to content ourselves with going to look at it once more (which this year entailed breaking in: the front door was covered in plastic and nailed shut, because it didn't have a handle or lock). The stairs were fixed, finally, but there was still no balustrade, so that you could fall sideways into the hall, or off the landing on to the limestone floor below. There were only carcasses for kitchen cupboards and none of the appliances was fixed. A leak appeared to be spreading across the sitting-room ceiling. The shower heads were in the wrong position: Freddie would have had to adopt a hunchback of Notre Dame position to wash his hair.

(Mavji had promised to move these shower heads a couple of weeks earlier: it was just a question of getting the tiler back, he said brightly, as if tilers were like obedient dogs and came when you called. But I don't think they can have been, because the shower heads stayed where they were. That it was simple to move them was just another thing they thought we wanted to hear, similar to Ramesh's instructing Mavji, 'Do it tomorrow!' after I'd noted that the children's bedrooms needed doorstops to protect their wardrobes from being bashed by bedroom-door handles. I mentioned this to Joyce a few days later: she said that this would require her to specify doorstops, so that they could then be ordered. There had never been any question of doing it tomorrow.)

I extended the lease on the house in Wapping for another fortnight. The new house was starting to feel like a mirage, slipping away the closer we came. All the way through I'd worried about what would happen if one of us died: would the other be able to carry on? Would it be possible, financially, not to? What if both of us died: what would the children do? (Though presumably, with a six- and a two-year-old to look after, a half-finished concrete house would have been reasonably low down everyone's list of priorities.) Now I worried that some catastrophe would occur at the last minute to stop us ever living there, to serve us right being so pleased that we'd got this far without being sent to prison for non-payment of taxes, for still being married. I don't know why I wasted quite so much energy on notional disasters since the lack of money was a real one, and quite enough to be going on with. But then Steve called back. He'd sorted it with one of his contacts. We could have the new money on the mortgage, pay the bills, move into the house. It was, he said, the very last thing he was doing at Barclays. 'We did the whole thing on a wing and a prayer, didn't we?' he observed cheerfully. We did; and it had made Barclays an enormous amount of money. If I'd been them, I wouldn't have let Steve anywhere near a redundancy package; I'd have nailed his feet to the floor.

On 6 January, I was able to inform Mavji that we planned – finally, definitely, regardless of whether they were ready – to move in on the 10th. It was obvious that the house wouldn't be anywhere near complete, but I thought I might well go mad if I had to spend another week in Wapping. I had to move on a Friday, because I couldn't afford to take any more time off work – I'd taken holiday to move the first time and had expected to be able to use the Christmas break to get straight.

Mavji looked alarmed, but bit his lip in a sign of determination.

We agreed that certain things needed to be done before then. My own aesthetic contribution to the project had been to choose the lights over the kitchen table. They were hideous; as soon as they went up I realized they'd have to come down. They had too much chrome (any chrome would probably have been too much, but they had a *lot*) and they looked designerish and affected.

(My efforts with the lights are probably best glossed over. After hunting for a new pendant to replace the oversized one I'd chosen for the corner of the sitting room, I was forced to acknowledge that the first light that Joyce and Ferhan had shown us – the Louis Poulson design, based on an artichoke – was the best I'd seen. I didn't know nearly as much as I thought I did about lighting, and what I did know wasn't tasteful.)

Meanwhile, in all the confusion over American or English ice dispensers and freezers, nobody, Mavji claimed, had ever told him that the fridge was now 60 centimetres wide rather than 90 centimetres (a size of fridge that, as we had discovered, doesn't exist anyway). That meant that wires which should have been hidden by the fridge sprouted instead from the middle of the limestone. There was some talk about patching in a piece, then about perhaps waiting for Joyce to come back from Florida and asking her what to do; but I realized this was pointless, because we all knew what Joyce would say. She wouldn't want any patching in. She'd want a proper limestone tile in that place, the same as everywhere else. And Charlie thought that Varbud ought to lift the tiles and replace them before we moved in, because the job would be so messy.

I told Mavji that there were three other things critical to our occupation of the house: a balustrade around the void at the top of the stairs, so that people didn't reach the top step and promptly fall 10 feet back to the ground floor; a working fridge, and a cooker.

I liked the idea of moving on the 10th. Charlie and I had met on the 10th (of the 10th, in fact) and we'd moved into our previous house on the 10th of August ten years earlier. Unfortunately, the removals men called on the morning of the 10th to say they were double-booked. The boss, they said, was about to ring us to ask if we'd mind moving the following day; they wanted us to refuse, because they didn't want to do the other job. But the boss was smarter than that: he claimed the van had broken down and he wasn't sure if he could get a mechanic.

It was probably just as well because when we got to the house late on the Friday morning, none of the three things had been done. Charlie and I went for a disgruntled lunch at the Vietnamese cultural centre in Hackney and called Ferhan to tell her we were furious: it was impossible; we'd taken the day off work to move and we couldn't ask the removals men to come in because they'd get to the top of the stairs, then fall and break their necks. (We didn't bother to mention they weren't coming anyway.)

Ferhan did a bit of shouting at Varbud and by the time we got back to the house after our bowls of noodles, Mavji was promising to have the balustrade done by 5 p.m. We decided that since mattresses had been delivered by John Lewis, we might as well retrieve the children from Wapping and spend the night in the house anyway.

Varbud got the balustrade up at 5.30 and I let the boys in. We had fish and chips and champagne and, a bit later, settled down on our new beds. There was nothing in the house but sheets and a few clothes: it felt perfectly, deliciously minimal. This was, or should have been, the optimum moment – for architects, the handover is often the beginning of the end. 'Very few of the houses', Frank Lloyd Wright once complained, 'were anything but painful to me after the clients moved in and, helplessly, dragged the horrors of the old order along with them.'

There was, I realized with delight, something liberating about this possession-free space, this quiet minimalism. There was room to breathe, somehow, in the absence of clutter. It was going to be all right after all, this stuffless life. Being here – I woke up to sunshine and an absolute conviction about it – was going to be good. All I needed to make life complete was the piano, the dining table and the books.

The removals van arrived. In came the piano, the dining table and the books, then box after box after box. What could be in them all? I had no idea we had so many possessions. How *could* we, after everything I'd thrown away? I'd only been separated from these things for a month, but I'd completely forgotten I had them. Where was I supposed to put them all?

I was still grappling with these questions a fortnight later, as I worked my way through the unpacking. Charlie lit fires in what would one day be the garden and I intermittently threw the entire contents of boxes on them, too exhausted for yet another trip to Oxfam. (One day, Neven caught me burning a paperback copy of *A Child Called It*, a book I dislike intensely and had only ever owned because I'd had to write about it. 'You should not burn books,' he reproached me gently. 'We are building a library at home in India. We will take anything.' I felt distressed: I was aware that burning books, historically, is only a small step up from burning people. All the same, I remained reluctant to inflict Dave Pelzer on Gujarat.)

Regardless of how many possessions I despatched, there still seemed to be a mountain more. Every moment I wasn't working or actively feeding people, I was battling with boxes. Maybe this was why I suffered in those first weeks from low-level depression; or maybe it was because after all the waiting, getting into the house was a bit of an anti-climax, just another list of jobs to do to the accompaniment of drilling and sawing and fixing from our now live-in companions, VM and the guys.

Maybe the depression derived from the realization, now that we were here, of how far we still had to go. Or maybe it was easier, once we were living in what was still basically a building site, to focus on how many things still weren't right. There was a drainpipe that came down through the kitchen: every time it rained, it sounded as though someone was playing a primitive musical instrument of a stones-inside-a-gourd kind, very loudly. And I worried ceaselessly about stupid, fixable things, about the white walls getting scuffed and why two of the blinds didn't work.

Possibly, though, I wasn't getting enough sleep. The burglar alarm kept going off in the middle of the night. VM had programmed it with Varbud's telephone number – only, unfortunately, not the one we used. When they did tell me the code, I managed to transpose three of the four digits in my head.

If I wasn't stabbing randomly at the burglar alarm panel to stop the screeching sound, I was being kept awake by flapping plastic around the edge of the roof. This plastic couldn't come down until the shutters went up, but the shutters hadn't arrived and no one was entirely sure that when they did they'd work anyway. Joyce and Ferhan had wanted simple metal frames supporting horizontal slats, but Brian had said that, at the size, they'd droop at the edges. Together they'd decided to incorporate a diagonal metal strut. A prototype was coming soon, but even Brian still couldn't be certain it wouldn't bend. 'That's the nightmare about engineering,' he said cheerfully: 'it's often very difficult to prove something, so you have to stick your neck out.'

Ned kept asking if we were going back to the old house and, sometimes, if we could return to Wapping. He was the only person who had really liked it there.

'The old house was nice,' he said obstinately over breakfast one morning.

'This house is nice,' I reassured him. 'And it'll be even nicer when it's finished.'

He looked at me disconsolately. 'But it's just an advert.'

It didn't feel like an advert; not, anyway, in the magazine architecture sense. Architects – some of them, anyway – have a habit of working towards the moment of handover as if that were all that mattered; as if the crucial thing were not how people adapt themselves to spaces, and spaces to people. 'You get work through getting awards, and the award system is based on photographs,' complains Clare Cooper Marcus, of Berkeley's

Architecture Department. 'Not use. Not context. Just purely visual photographs taken before people start using the building.' The British architect Frank Duffy similarly speaks scathingly about 'the absolutely lifeless picture that takes time out of architecture – the photograph taken the day before move-in. That's what you get awards for, that's what you make a career based on. All those lovely but empty stills of uninhabited and uninhabitable spaces have squeezed more life out of architecture than perhaps any other single factor.'

Any photograph of our house taken the day before move-in would have featured a gang of Gujaratis on a building site. And still, weeks in, it seemed inconceivable that the house would ever be finished, that groups of Varbud operatives wouldn't be turning up in the morning during breakfast, waving through the windows as they went off for their first Indian tea, that I would ever feel calm.

In the past, when people had claimed that moving was the third most stressful life experience, I'd thought, yeah, sure, moving is *really* in the same league as death and divorce. Now I had fungus on my feet and an ulcer on my lip. I'd lost one shoe from a pair I'd bought in New York, a mobile phone I'd only acquired the previous week, and two accounts books with all my figures in them for the last four years.

I worried about terrorism, and then that I was becoming like my mother, who used to push her pram around Leytonstone thinking apocalyptic thoughts, looking at other people and wondering if they were thinking the same things. Admittedly she was raising children through the Cuban Missile Crisis and the Cold War and she was probably pretty depressed about being at home all the time. But I know how she felt. And I suppose I thought that if I could envisage things, then they might not happen – like an amulet, almost – because it seems to me that it's the things you don't think of that mostly happen. I envied Charlie his absence of a sense that life is precarious, and then I felt bemused by it because life *is* precarious, and if you've had one of your parents die young and unexpectedly, how can you not be left with a sense of imminent dissolution?

It took me about three weeks, but then I stopped being depressed. I drifted through my now very organized house, where the stuff that was worth keeping was put away in cupboards, hidden so tidily behind slabs of iroko that it might as well not have been there. The spaces seemed spacious, breathable, open to the sky. When Charlie had insisted that the house would feel big enough, even though it wasn't actually very big at all, I hadn't really believed him. But he was right: there was the impression of room to move. And the sun seemed to shine every day, rather as it had the autumn that Charlie and I met.

Having other people come to visit helped. Their presence seemed to ratify it as a proper house, a place where I could serve dinner and from which I could send them away again at the end of the evening. It didn't particularly matter what they thought of it. My friend Robin looked around sceptically and said he could see that it all fitted together, but he was glad he hadn't had to build a whole house: he'd found it horrifying enough just needing some new taps for the bathroom and realizing there were 176 options. He'd felt alarmed by the degree to which his identity seemed to be caught up in the bathroom tap issue and, for a whole house, he wouldn't have known where to begin – unless, he added, he were building an ecological house, which would give him a theoretical framework. (He was saying, I think, that he wouldn't have been ready to sign up to the persistent faith in modernism that our house represented: for him, as a set of principles, it just wasn't compelling enough.)

My friend Lindy was more straightforward: she could see the point of it, she said, but she couldn't have lived in it – too many straight lines and hard angles.

Hugo, who was often in London, visited several times in the

early weeks, and kept describing it as witty, which I couldn't understand at all. What did he mean, 'witty'? Was that a good thing? Witty sounded alarmingly postmodern; jokey, like a gingerbread house. Witty was something you looked for in people and novels, not the place you had to live.

But most people seemed to like it. (Actually, Hugo loved it too. He just thought it was witty.) And anyway, I didn't care whether they did or not. At a practical, day-to-day level, it made me happy – sitting on the kitchen bench looking out into what would one day be the courtyard-garden, getting into bed in our sleeping cave of a bedroom, the view from my study along the concrete wall, squirrels rootling around and skipping through the slug garden.

And we did have a few magazine architecture moments: they just came later. Joyce and Ferhan gave a reception at the house one Wednesday afternoon in March for clients and friends. This proved to be a very good idea of ours (not that this had factored in our initial suggestion) because it imposed a deadline that Joyce and Ferhan were absolutely determined to meet. It meant, for example, that the day before the party, the shutters arrived. (At some point, the original plan to have roof-to-ground shutters along different parts of the façade had shifted to shutters along the entire upper storey.) And, as they went up, the morning of Joyce and Ferhan's event, they transformed the look and the feel of the house.

Downstairs, the house felt open and fluid, as if it were teetering on the verge of outside; upstairs, it was screened, protected, intimate. The (fixed) slats were 10 centimetres apart, wide enough to be able to see out clearly, but narrow enough to obscure the view in from the lane. The shutters opened out to nearly 90 degrees, to let in spring and autumn light. Closed, they cast stripes across the concrete wall in our bedroom and dappled the floor.

The builders worked more frantically in the run-up to the

party than at any other point in the process – hanging wooden doors on the garage cupboards across the courtyard, putting up the handrail on the stairs, fixing the mirror and the lights in our bathroom. Joyce and Ferhan sent a window cleaner (until then there hadn't been much point, because the builders were making such a mess). He spent all day washing and polishing. The following day, the day of their party, they arrived with armfuls of flowers and moved the new lamp I'd brought into a cupboard, replacing it with the one they preferred.

The day after that, while everything was still looking pristine, a photographer came and took pictures. Perhaps mindful of the hostility to sterile, magazine architecture shots, he left Ned's wooden truck artfully in the middle of the sitting-room floor. The pictures were featured on the front cover of the *RIBA Journal* and the *Architect's Journal*. We won an RIBA Award and were shortlisted for Best House of the Year and Best First New Build.

The builders, though, still weren't finished. They didn't leave until May, although work slowed for a while in spring when VM took three weeks off to go to Gujarat. A temple that had been destroyed in the 1998 earthquake had been rebuilt and was being consecrated. VM explained that he had taken it upon himself to be in charge of the cooking. We asked politely how many people were coming; he said between thirty and forty thousand.

Joyce was often around as well, discussing ironmongery (for some reason the wrong locks were always being ordered or sent) and absent-mindedly pushing my tealights up the bench, or moving the one photograph we had dared to put on display from the middle of the piano to the end.

Occasionally, aspiring self-builders got in touch to ask if they could come round too. They patrolled the spaces asking detailed questions about prices per square foot (the answers to which I almost certainly got wrong) and planning applications. They didn't seem to feel there was anything *de trop* about commenting adversely on things we'd done ('Hmm,' said one woman, opening a cupboard door without invitation, 'this sticks out a long way'). They knew far more about building a house than we did; like members of a cult, they were possessed of arcane knowledge about orientation on the site, ceiling heights and varieties of limestone. I told them some of my little tales of ineptitude and felt I hadn't really supplied whatever it was they wanted – which would have been difficult because, in most cases, it was a site. Some of them had been looking for more than five years. I wasn't sure I could have endured their successive, exhausting, failed attempts to secure patches of ground. It struck me that they must have enjoyed the troubled state of mind this process induced; there must have been something about being in the self-build cult that was rewarding in itself, regardless of whether any buildings were actually going up.

It became apparent, as they talked, that the huge expansion of conservation areas in recent years has been used by timid local authorities as an excuse to stop all new building, even of a potentially interesting and sympathetic kind. We'd been luckier than we'd realized that our site had been in such an out-of-the-way, scruffy corner, where the architecture had no consistency; lucky, too, that Islington Council is well disposed towards new architecture. In many local authorities, would-be new builders are regarded as a species of vandal, to be obstructed at all costs. Charlie said he thought that in addition to conservation areas, there should be new-build areas, where people had to knock down all but the very best houses every fifty years and put up new ones.

We were asked to take part in London Open House, the annual opening of buildings of presumed architectural interest in the capital, which takes place over a weekend in September. We could have opened all of the Saturday and Sunday if we'd wanted, but, since we needed all of Saturday to tidy up and anoint the stainless steel with baby oil, we confined ourselves to Sunday afternoon between 1 p.m. and 5 p.m. I expected about sixty people. That started to look unlikely when sixty turned up on the Saturday. The programme quite clearly stipulated our opening times, so I don't know why people were so disgruntled when we said we were closed. But since they'd made the effort to come, I let them look round the outside (which, being mostly glass, wasn't that different from looking round the inside). Some wandered in anyway. (It was a nice day and we had the odd door open.) I found one bloke sitting on the kitchen bench with a plastic bag on his lap.

'I'm terribly sorry,' I said, 'if you want to look round inside you'll have to come back tomorrow.'

'Oh no,' he answered disparagingly, looking at the concrete walls, 'I've seen *quite* enough.'

By noon the following day, the queue was halfway up the lane. I'd stipulated in the programme that we'd let in six people at a time, thinking that otherwise they wouldn't be able to see anything. But though we opened half an hour early, it rapidly became apparent that this six-people thing wasn't working. We began admitting fifteen, then twenty, then thirty, keeping them by the front door while we gave them a little spiel transparently designed to hold them in one place while the previous group made their way round and, we hoped, got out. (Most people were cooperative about this; a few lurked for hours.) Ferhan had come to help (Joyce had opened her own house, so couldn't be there), and she initially gave this speech, since she knew what she was talking about; but she had a sore throat before she started, so then Charlie and I had to take over, when it became not merely an obvious delaying tactic, but also architecturally ignorant. (Many of the visitors were architects or architecture students, and should really have been giving the speech to us.)

Many of the visitors had very specific interests.

'How is the roof constructed?' one woman demanded.

'Er . . . flat.'

'Yes, but how is it *constructed*?'

Another person was exclusively concerned with the Louis Poulson artichoke light fitting. I pushed past the crowds coming the other way, to go downstairs and switch it on for him, but then he wanted to have a conversation about voltage.

At 4.45 I trailed up the lane and asked the last person in the queue not to let anyone join on behind her because we quite obviously wouldn't be able to let them in. When I went back half an hour later, the line stretched as far up the lane as ever.

'I'm terribly sorry,' the woman said, almost tearfully, 'but they started *shouting* at me.'

I yelled up the lane that we'd only go on until we reached this woman or until 6 o'clock, whichever was the later. I got a bit muddled with this concept – whether I meant later or earlier – because I hadn't eaten anything since breakfast or even had a cup of tea. (It seems wrong to make yourself a cup of tea in front of people who have been queueing for two hours.) Still, quite a few people got the message and peeled off the line.

We let in everyone who stayed. We more or less gave up on the speech at the end, which probably made it a better experience for the visitors, even if they were squeezing into the bedroom like illicit lovers in a broom cupboard. My mum stood by the back gate, getting people to write their names and comments, if they wanted, in a little book. (London Open House suggested this might be a good idea.) 'Huge small spaces,' wrote someone, which was about right. 'Raw, calm and unique,' said someone else. Lots of people loved the bathroom and kitchen and a number said to let them know if we wanted to move.

Despite the wait, the atmosphere was enormously good-humoured: the English middle classes doing two of the things they like best, queueing and thinking about houses. 'Very much enjoyed talking to the teenage son and hearing his version of events,' one person wrote ominously.

We finally turfed everyone out at 7 p.m., only a couple of hours late. The intensity of people's interests, and the perfect weather, made me think that, after all, we had done something special. In the early evening, with the children standing on the wall, my mum chatting to my Auntie Evelyn, who'd come up with her friends from Bromley, sunlight still seeping from the courtyard into my kitchen and several people sunning themselves in the garden who, I'm sure, had been there all day, rather as if we were a picnic spot, I felt washed by happiness. 'No dreamer ever remains indifferent for long to a picture of a house,' Gaston Bachelard said. Our particular dream had become a place in which other people had now dreamt, and where, as we gradually accommodated ourselves, we might start having new dreams of our own.

Of all the comments in the books, my favourite was 'Un concept et une réalisation parfaite; au niveau de la qualité des matériaux, de la lumìere, de la circulation,' although possibly just because it was in French. A man from New York wrote: 'Such an extraordinary statement as a home – it must be difficult to live in at times.' The thing is, he was completely wrong.

Since the house was always conceived of as being seamless with the outside, we were keen to get the garden started. I interviewed two garden designers – or, strictly speaking, one garden designer and one landscape architect. The designer was more concerned to find out what kind of planting I liked (and I think would have given me more leeway) but started talking alarmingly about painted walls and terracotta, either of which would have caused Joyce and Ferhan to have seizures. The landscape architects had worked with Azman Owens before and would, I felt, be more likely to complete the job with everyone still on terms of reasonable civility.

In fact, Joyce and Ferhan had plotted out the garden very early on and the basic structure didn't change: a limestone bench on a low concrete wall by the ash tree, a table and chairs on a limestone terrace, a water feature. There wasn't that much flexibility about all this – the table had to go over here, the path there – on account of the views from the house and the traffic from the house to Charlie's study. The landscape architects, del Buono Gazerwitz, seemed to understand this and regard it as less of an imposition.

My only real reservation about Joyce and Ferhan's design was that it didn't contain enough planting. I didn't think I wanted the lawn they'd suggested (the space would be too small for really satisfactory games of football or cricket even if we grassed the entire thing. My view was that the boys would just have to make do with playing outside in the lane, or going up to Highbury Fields, which was, after all, only seconds away). And I was anxious that the garden should complement the house, rather than slavishly follow it: I wanted something that was a counterpoint, rather than a reconstruction in organic matter. Not, in other words, a completely modern garden.

While I was thinking about what kind of garden I *did* want, I visited the Contemporary Gardens Festival at Westonbirt in Gloucestershire, which features a series of cutting-edge gardens. My favourite was the Cement Garden, which, since it was 'inspired by building sites', we pretty much had already. Instead of trees, it featured a grove of hard hats stuck on metal poles. Reinforcing rods of the kind used inside our concrete (and still lying around plentifully) were twisted to make a 'hedge'. And there were some straggly plants that looked like weeds, mainly because that is what they were – e.g. cow parsley and dandelion. Not too many, obviously, which would have undermined the building-site atmosphere.

An avant-garden like this would have required very little work; we would have dispensed with the expensive services of landscape architects and garden construction people. But it wasn't quite what I was after. (Modernism, despite the best efforts of garden designers in the last couple of decades, has never really got hold of gardens in the way that it has buildings. What passes for modern often looks back to ancient Japanese, or the medieval *hortus conclusus*.)

I tried the Chelsea Flower Show instead, where the winning garden was designed to look undesigned, more like a wildflower meadow, and again featured some rather weed-like plants, artlessly disarranged. The meadow look has been fashionable for several years now but unfortunately isn't nearly as casual and effortless as it's made to seem. It's bloody hard work, because many of the favoured plants have thuggish weed-like habits and need pulling out, dividing and restraining. Since we are only intermittent gardeners, that was out for us too.

Snobbery and fashion are integral to gardening: it is perfectly possible that Nebuchadnezzar looked down his nose at other Babylonians whose gardens didn't *hang*, and medieval monks sniggered over one another's herbs; but, for certain, snobbery has been vital to gardeners, at least since they came over all romantic in the eighteenth century. 'Nature abhors a straight line,' declared the designer William Kent, presumably after watching apples fall in that wiggly way they have.

This makes designing a garden from scratch very daunting. You probably don't want to align yourself with the suburban bungalow lot even if you own a suburban bungalow – which, obviously, we didn't. (Brightly coloured bedding has been deplored by the sophisticated at least since Ruskin talked of garden plants 'corrupted by evil communication into speckled and inharmonious colours'; nowadays, begonias in garden centres come in improbable hues of a kind otherwise only seen on dress and coat ensembles worn by the Queen.) On the other hand, you don't want to be a faddish designer drone either, the sort of person who buys olive trees because they're a big look right now and puts them on a cruelly exposed balcony where they promptly die.

The clever thing, clearly, would be to pick and choose, creating our own style, reflecting our personality and honouring the spirit of the site along the way. But unless you're a plantsperson, that's tricky. Like surfers in search of the perfect wave, plantspeople are engaged in a perpetual quest for exactly the right place for their plant. They only really like species plants – i.e., ones that once grew in the wild – and they don't, on the whole, favour colours, interesting themselves primarily in texture, shape and form – unless, that is, they're followers of Christopher Lloyd, who has created a painterly 'hot garden' at Great Dixter in Sussex. Lloyd's admirers like exotic plants and anything orange, and in that sense they're like the suburban bungalow lot, except of course completely different . . . One way and another, creating a garden from scratch was strewn with possibilities for horticultural faux pas.

Charlie and I once visited the town of Saint-Rémy, in Provence, where there was a square we especially liked: paved, gently sloping, surrounded on three sides by tall buildings, with a tree and a bench: plain and, in many ways, unexceptional. But on the upper side, it featured a kind of standpipe (maybe there was an underground spring) out of which water trickled into a little trough at the base and, from there, into a rill that carried it all the way round the edge of the square. We loved this place, this open drain, and Charlie said we should keep it in mind when we did the garden. What we ended up with looks nothing like it, but I like to think still echoes that square, in the sense that we do have water running from one pool by the bench into two others in the slug garden. And increasingly, we started thinking of the garden as a courtyard.

Tommaso del Buono, who drew up the designs, initially wanted grass, or at least a camomile lawn, but we eventually decided (Joyce, far from siding automatically with her fellow professionals, as I'd feared, backed us up here) that this might make the garden too bitty. We would have a slab of planting in front of the den, kitchen and a bit of the sitting room. I was worried that the basic design and hard landscaping were too formal, but Tommaso promised me that planting in drifts would mitigate the effect of the straight lines and serried rows of limestone tiles.

The scheme he came up with had a slightly Mediterranean feel – lavenders, rosemarys, santolinas, catmint, salvias and creeping thyme – without making too much of a statement. The house makes enough of a statement as it is, and he also included grasses, alliums, peonies and poppies. The idea was for quite a lot of the plants to carry on looking good – since we had to see them all the time – through the winter. Round the back, in the slug garden, Tommaso planted mainly springtime white flowers, hellebores and *Gallium odoratum*, *Tiarella cordifolia* and some climbers (a white winter-flowering clematis, scented *Trachelospermum jasminoides*) and three Japanese maples. Around the ash tree, he chose plants with big leaves (anemones, brunnera, angelica). Against the concrete wall, we would have three large pots containing strawberry trees, and on the opposite wall, pleached pears.

Even people who were equivocal about the house liked the garden. But gardens, even more than houses, are always incomplete. They can't be works of art, despite what garden designers might secretly wish, because they are restless with sex and death. Architects work in three dimensions; garden designers – good ones, anyway – also have to deal more pressingly with a fourth, time. Things rot, set seed, or fail to, get aggressive. Only for a moment can they ever resemble the photographs you see in magazines (the pictures are usually taken at dawn anyway, when normal people aren't up, because that's when the light is best) with covetable plants ideally organized. I loved the garden and, particularly, I loved the way it lent the house a whole other dimension, but I was under no illusions. Weeds would sprout. Everything, in due course, would die. (This is perhaps why we love gardens: they're like us.) I was not on top of this garden and

I never would be, wholly: from now on we would be locked in struggle. The partialness of our control over these trees and flowers was what made them fascinating, and beautiful.

Initially, much of the pleasure we experienced at being in the house was sheer relief. Bills were still coming in, but not with such whirlwind force or in such unpredictable amounts, and we felt less battered by them. Besides, we finally had something to show for it all.

It was also a pleasure to be less dull. Over the preceding months, we'd become house-building bores – which was just about excusable when the project was fun, but tiresome once it became a grind. And the grind part had gone on for some time: it was easy, in the thick of it, to forget how long the process had taken, but over the course of it, the shape of our family had shifted.

Henrietta, for a start, wasn't around as much any more. I was slightly embarrassed by the fact that no sooner had she decided to settle with us than we'd announced that we intended to be off ourselves. We'd waited all those years for her to stop flitting about, and as soon as she did, we decided on a flit of our own. Ferhan often said to me – especially when I worried about the size of Hen's bedroom – that I shouldn't worry so much, she'd be gone soon anyway. But I didn't want to think like that. Aware of the fragility of her feelings about home, I'd always wanted, still wanted, the house to be about her as much as anyone.

Inevitably, she'd been more removed from the process than anyone else: she went off on her gap year before we started on site, and by the time she came back, we already had half a house. We moved in after she'd gone back to university for the spring term and she didn't spend a night in the house until weeks after the rest of us. But she talks easily about the house as home and seems relieved to get back at the end of term, more at ease with her multi-homeness than she's ever been. She quite likes it that she occasionally meets people in the pub who say, 'Oh, *you're* the girl with the amazing house.' And her numerous friends visit, which pleases me because I always wanted to have people coming and going. (Luckily, we haven't found that when a new one comes into the kitchen, another has to leave.)

Freddie's requirements had changed even more noticeably during the long process. Back in the design phase, he'd been mainly preoccupied by stick insects, praying mantids and salamanders. Joyce and Ferhan had designed a glass box to protrude out of his room over the stairs, so that everyone could have the pleasure of these creatures. (They liked this slot: they used it for their Christmas card that year.) By the time we moved in, two and a half years later, the creatures had gone – the stick insects into the curtains of our previous house, the salamanders, sadly, to something called salamander rot, one praying mantis at the hands of a murderous mother. Freddie was now mainly preoccupied with privacy. He hung curtains over the back of the glass box and grumbled about noise filtering through it when he was trying to sleep late on Sunday mornings. The box became a display case for deodorants.

He remained doggedly sceptical and scathing to the end, refusing to throw away anything before we moved, declaring that all his possessions were crucial to his well-being. When we'd been in the new house about a month, he appeared from his room with around a dozen plastic sacks of rubbish, explaining that there were, after all, a number of things he could no longer see the point of. He did manage to salvage one of the Persian carpets from the move, along with an Arts and Crafts book table and a battered table lamp, so his room has a slightly different feel to any of the others. But not that different, because Freddie principally likes things to be clean and comfortable. (He says, for example, that the utility room is one of his favourite parts of the house, because it means I no longer leave piles of washing on the kitchen floor.) Despite his hostility, he was the first to bring friends round. He still grumbles about the noise from not having carpets and the modernist aesthetic generally (occasionally threatening to paint the concrete in his bedroom blue) but mainly, I think, because he believes we'd be disappointed if he didn't.

While we were designing the garden, we looked, finally, into the possibility of incorporating a tree-house, although the brief did not include either a telephone or television. For some reason, the task was entrusted to a sort of artist in natural materials, a sub-Richard Long figure, who came up, first, with a sort of woven nest (dark, prickly in your back and bottom, nowhere to put down your cake and apple juice) and, second, with a platform bounded by scaffolding poles (horribly visible to parents). Joyce didn't like either of these, and neither, on reflection, did we. The house was too small and too coherent to start introducing strange, egotistical structures. It became increasingly apparent that any tree-house would probably have to be designed by Azman Owens and made of iroko. But the garden construction came in over budget (for some reason, I did not find this surprising), to the extent that, even by our second summer of occupation, we hadn't managed to afford a table for outside eating. It is quite possible that by the time we can afford a tree-house, not only Harry, but Ned, will have outgrown the whole thing.

Not that the garden is without its pleasures for Harry. The year after we moved in, he developed a passion for Arsenal. If you stand in our courtyard, you can hear the roar of the crowd when the home team scores at Highbury (that amazing season, this happened all the time). Even I find this moving. It must be quite something for an eight-year-old dressed from head to toe in home strip.

The rest of the house also has its attractions for small boys. The concrete turns out to provide a very good backdrop for photographs of Thierry Henry, Robert Pires and Jose Antonio Reyes cut out of newspapers. People (architectural types mainly) sometimes ask me if I think the children will develop a heightened aesthetic as a result of growing up in a sleek concrete structure. So far, their education in modernism has issued mainly in the liberal Sellotaping to walls of grainy and wonkily cut-out pictures of men performing 'skills'.

As we spotted when the house was pegged out on the ground, it's not big. Considering how many bedrooms and studies there are, it is really remarkably compact. Yet it never feels small; on the contrary, I wander around it feeling the opposite of hemmed in, as if there's room for anything. One reason why this should be so, of course, is that our initial reason for moving – all that baby stuff – had already gone by the time the house was finished. We took no buggies or bouncing chairs, nappies or beakers; by the time we moved, Ned was drinking out of glasses and using his legs like normal people. We didn't even take his cot, deciding that he could start his new modern life as he meant to go on with it, in a new modern bed. (In fact, he actually started it on a mattress on the floor, because it was several months before we could afford to get the beds built.)

Another, more positive, reason for the feeling of spaciousness is that all the space is working for us: there are no rooms, or corners of rooms, that we don't quite know what to do with, to become wasted, untidy, irritant places. The glass, combined with the wraparound garden, gives a sense of openness (literally, in the summer, when the doors go back and house and courtyard are absorbed into each other). And our freedom of movement, or impression of it, is enormously helped by the absence of clutter. The wall between the hall and the kitchen isn't actually a wall at all, but a length of deep cupboards, with the sink, cooker, fridge and worksurface on one side, and masses of storage space on the other.

These cupboards in the hall are big enough, and flexibly organized enough, to hold all the toys (mostly in big plastic boxes) as well as the food processor, spare pots and pans, box of fuses and plugs and screwdrivers, wine, cookery books, vacuum cleaner and pretty much anything else that we need to keep in the house.

Slightly to my surprise, I don't find it unreasonable to ask that only a couple of the big toy boxes should be out at any one time; still more surprisingly, the children don't seem to find it unreasonable either. We have attained what had seemed to me previously an ideal but improbable condition, in which toys are accessible without actually being in piles on the floor. The children seem to prefer their Lego in a separate box from the train set, on the whole, and the men apart from the music instruments. And they like the fact that the toys are all tucked away, so that they can forget they've got things and come back to them.

The real test will be whether it stays like this. It's easy to have everything in boxes the day you move in, but new toys will inevitably come, and space for them must be found. I have learnt something, though: it's simply not worth hanging on to old toys that aren't played with, or still worse, half-toys. I am more ruthless than I used to be. I think of myself as Binwoman – a sort of superheroine of decluttering (which is ironic, really, given that a bin is the one thing we failed to get). I am particularly cruel in the matter of party bag toys – admittedly, here, on strict instructions from Joyce. They hang around for twenty-four hours and, unless the children have proved themselves particularly attached to them, they go.

While toys have to be admitted, this is not the case with other stuff. We don't need any more furniture. At some point we'll want a bigger dining table, and we still haven't got a reading chair for the corner of the sitting room, but we've known about both of those since the day we moved in. Essentially, the house is complete. It's very restful to be free of consumer pressures in this fashion: there's no point in buying a Provençal plate in a French market, however buttery yellow and lovely, because there's no place in the house for the non-white. There's no question of seeing an unusual, stylish lamp and impulse-buying, because the lighting's already sorted (besides which, of course, I'm useless at lighting).

The house has become what we wanted it to be: a backdrop – a powerful one, in the sense that it makes us feel good – but one that isn't so insistent as to be distracting. Which is not to say that everything's perfect, or that there have been no teething troubles. One morning in September, eight months after we'd moved in, I looked up from my desk and realized the shelves above it were supposed to be a cupboard. I called Joyce, she checked the plans and, sure enough, Varbud had forgotten to put a door on. None of us had ever noticed.

For months it was actively painful to shower: face, neck and back were stung by tiny, fierce needles of water. Joyce and Ferhan shook their heads and said they used these shower attachments all the time and no one else had ever complained of similarities to torture by exotic martial regimes. In the end, someone fixed the problem by altering the water pressure.

The door to the bin store warped and didn't close for a while, so it banged on windy nights. The right ironmongery didn't arrive for the long windows at either end of the upstairs hall, so that they didn't open and made the hall stuffy all through our first long, hot summer. The concrete stairs showed an alarming tendency to chip and flake on the leading edges and needed abrading and recoating: everyone had overlooked the fact that untreated concrete crumbles where it isn't reinforced, especially when several boys are constantly running up and down it. And, eighteen months on, we also have yet to use the garage. For reasons that seem to change every time someone else looks at it (but which currently seems to be something about being too heavy for the mechanism) the door doesn't work. The one good thing about this is that it makes the purchase of the world's most expensive parking space from Tom Tasou seem slightly less humiliating. In the meantime, the garage proves to be an excellent place for storing my many pointless files of notes for old articles.

We couldn't get the hang of the underfloor heating: it kept melting chocolate and making raisins go hard. The individual room thermostats, which had seemed such a good idea, bore no relation to one another (some turned up to maximum didn't do anything; others made rooms swelter on the first notch). Things improved when we grasped that you had to adjust the thermostat twenty-four hours before you wanted to feel the effect (though some rooms – they tend to take it in turns – still simply tune out). You just have to hope the weather doesn't do anything sudden.

Still, compared to the pleasures, the inconveniences are minor. There are lots of specific things that, it turns out, were a design triumph, such as the waste disposal unit that Joyce insisted on, to universal scepticism, but which, as she promised, none of us would now be without; and the large coats cupboard by the front door, which means that everything from outside gets dumped as soon as people come in. Charlie says he thinks of the house rather as the Japanese do their properties – i.e., as if the whole place has the status of a bedroom, requiring bare feet and relaxed clothes. (He has developed a habit of throwing off his clothes and putting on jogging bottoms, which I'm not sure is a lifestyle improvement.) The dressing room is wonderful, too, because it means there are never any stray T-shirts or thrown-off jeans in the bedroom. There's nothing in the bedroom, in fact, except the bed (and a couple of drawers, incorporated into the bed design), which seems very soothing after years of having to skirt around piles of books with irritating names like *Funky Business*.

The loss, to budget cuts, of the outdoor shower space is Charlie's one regret. There is a non-opening rooflight above the bath and shower, and it's very nice indeed to shower on spring mornings with clouds scudding above, or to sit in the bath on a summer night, stars scattered across the sky. And I am sure he's right that (on a few days a year, anyway) it would probably be even lovelier if the glass above retracted. Joyce tells us it would be easy enough to get this done, but this presupposes a recovery in our finances to the point where we can contemplate lump sums.

Our bathroom has been such a success generally that it's difficult to get Harry and Ned out of it. They have their own bathroom – a fetishists' delight of stainless steel and red rubber – but retain their enthusiasm for what they call the Japanese bath. This is deep enough for them to be at serious risk of drowning, which they much prefer.

The den is as un-denlike as I suspected it would be, not least because it's the room that most persistently refuses to respond to the underfloor heating. So Ned and I don't sit on the floor doing jigsaws, as I had once fondly imagined, although Harry has quietly commandeered the space and invented a new role for it as the X Box Room. Ned prefers running round and round the interconnecting rooms of the ground floor to doing jigsaws anyway.

I'd like to be able to say that the kitchen had enabled me to cook like Hugo. A veterinary nurse who once came in to feed the cat did ask me: 'Is your husband a professional chef? Only I thought, seeing your kitchen . . .' Why she should have thought Charlie was the cook, I don't know. The truth is that neither of us is great in the kitchen, although now that we have somewhere to put pans down when we get them out of the oven, our slightly chaotic methods are at least less likely to result in serious accidents. Despite the kitchen's failure to have any noticeable effect on my culinary performance, I remain delighted with it, because it's highly functional, but also, cleverly, looks as though no one does much cooking in it. In my more charitable moments, I'm even prepared to admit that if we had a bigger fridge, bits of food would probably get left at the back and start smelling.

We've succeeded in becoming a family that only owns white

mugs and two basic shapes of glasses, although our lifestyle changes have, disappointingly, not extended to my wearing Armani. But the floors, at least, are elegant: if I did an inventory of this house similar to the one I did at Malvern Road, of things-out-of-place-mainly-on-the-floor, I wouldn't be able to come up with much more than one handbag and one bin bag (which doesn't really count anyway, as it's the bin). I can't really understand why being this tidy doesn't seem a strain, and can only conclude that it was never my fault in the first place. I was conspired against by self-messing environments.

I do have a cupboard in which I shove Charlie's post. I wouldn't dare throw any of it away, but I do occasionally cart it across the garden to his immaculate study and dump it on the desk. In summer, Charlie opens up the glass study wall, takes his chair round to the other side of his desk, and works on the decking, underneath the ash tree. My own study is not nearly as beautifully organized as his, although I finally have enough shelves not to have to use the floor as a filing space, which for me is a major step forward. Some of the judges who came round to assess us for one of the awards exclaimed at the delightfulness of my study, which surprised me, because I think of it as a funny little room at the back of the house. It too, though, turns out to be rather cleverly designed: just sufficiently cut off from the rest of the house for me to be able to work undisturbed, and near enough to be aware of what's going on, to be visited by Ned on his way in or out. Besides which, it has a wall of glass, out of which I look on to spring flowers, two pools of water, and a grove of Japanese maples. (I look out, in fact, an awful lot.)

The slug garden, then, turns out to be a great success. The best thing, though, is not the rooflight over the stairs, or the cupboard space, or the benches, but a kind of cumulative effect of all of them. The house has an atmosphere – cool, calming, elegant, light, open – that is hard to pin down, but which you experience the entire time you're in it. Sometimes it seems to do with the glow of the floorlights, at others with the view from the bench across the kitchen into the courtyard, up into the branches of the ash tree; sometimes with the abstract interplay of materials and light. I couldn't understand this at first – despite having invested all this time and money, I was still sceptical about the extent to which living in the house could affect me (or perhaps should affect me: after all, I was happy before, wasn't I? Surely I should be fine as long as Charlie and the children and the rest of my family were OK?). But then a friend suggested that it was rather like the difference between waking up on a cloudy morning and a sunny one, and it seemed to make more sense.

In spite of the concrete, and the clear floors, the house is unquestionably a family house. It's warm, and full of toys and footballs and cricket bats and CDs, of Christmas trees and candles in winter, and sprawling teenagers barbecueing in the courtyard in summer. It doesn't feel like it belongs in a magazine. But it's also sleek and calming; a house where you don't get snagged on things out of place. The house is all of a piece, and it makes us feel all of a piece.

Postscript

There is a term on the fringes of biology, 'ecopoeisis', which refers to the process of a system making a home for itself. The process of occupation would be even longer than that of creation, and in itself creative: our task now was to find rhythms, routines, to start to feel settled in this concrete house. Charlie displayed radical ecopoeitism by getting up first every morning. It's difficult to convey what an extreme adaptation to environment this was. When I met Charlie, he habitually got out of bed somewhere towards noon; he may have become a journalist precisely because it was one of the few jobs in which you could still do that (sadly, it's different now). For as long as we'd lived together, he'd almost never got out of bed before me. After we moved, he did it every day. He said he liked being the first downstairs.

We shifted our behaviour in other ways in deference to the house: brought in the white mugs, took off our shoes at the front door in wet weather. But ecopoeisis isn't, presumably, an entirely one-way process, and we also waited to see how the house would wrap itself around us.

In the past, putting up pictures had been one of the principal and most urgent ways in which I laid claim to any new place. This time we didn't do it. There was, admittedly, a practical consideration: hanging things on the concrete walls would have meant puncturing them, so it seemed pretty important to be sure they were the right things. What we really needed, I kept saying wistfully to Charlie, was a Chris Offili. Needless to say, we didn't have a Chris Offili or any prospect of affording one. But there was also a sense of not wanting to rush in, heedlessly imposing ourselves. We'll probably put up some art some day. But for the time being, we're happy to look at the multi-hued, light-shifting concrete.

As any three-year-old exposed to the tale of the three little pigs could tell you, houses are supposed to last, which means that any judgement of them at the moment of move-in is bound to be partial and suspect. The real test of a building's success is what it is like to inhabit, not just for the first few weeks, but over years. This was a house that had been built for us, around us, but on the basis of a view of us that was, at best, a snapshot. We were already a different family – both, I liked to think, more diffuse and more close-knit – to the one which had started out on this project. The real test of the house would be whether it could absorb our changes – children leaving and coming back, everyone getting older, maybe grandchildren eventually, perhaps even (though it looked improbable) a time when we weren't working flat out to pay off the mortgage. The few magazine architecture moments we'd had, the house apparently suspended in a state of poised perfection, had been gratifying, but they weren't what living here was really all about. 'The art of living in its entirety – that is, the art of loving and dreaming, of suffering and dying – makes each life unique,' Stewart Brand has written. 'Therefore, the Cartesian, three-dimensional space into which the architect builds and the vernacular space which dwelling brings into existence, constitute different classes of space.' We could only hope that the two things would slowly come together.

It took us more than a year to get anywhere close to straight financially: cheques were still bouncing in November. I often wondered whether we might have done things differently: avoided coming so close to financial meltdown, got the builders to move faster; whether we'd actually learnt anything from our three years of house-building.

In retrospect, we probably should have trusted Joyce more, because she'd invariably been right. At the same time, I had a paradoxical sense that we should somehow have stayed more on top of the project. Much of the time, I felt I'd been in a miasma – about timings, money, what kind of a building we were getting, why we were doing it. But to have been more on top of it would have entailed a more confrontational, combative relationship with Joyce. Charlie and I don't really work like that (not least because we're not much good at it). And I'm not sure that in the end we would have ended up any happier. The builders overran, but that's what builders do. And Joyce delivered the house at a budget of £540,000. A year on, that feels like a bargain.

That autumn, Joyce, told us she was leaving Azman Owens, and the country, and going back to Florida. It felt like very bad timing: we'd only just stopped being her clients and started being her friends and, as a friend, she was confiding and funny, mischievous, interesting, energetic. In reality, of course, her timing was extremely good: she'd been wanting to leave for ages and had only hung on so long to see the house through to completion. (Her mother had been calling at least once a month for the past year to find out if she was ready to come back, and Joyce would plaintively reply that she still had to finish the concrete house. I think it's fair to say that when her mother eventually saw it, she couldn't quite grasp what all the fuss had been about.) We were almost the last people to know she was leaving, because she hadn't wanted to add to our anxiety.

But it also felt like bad timing professionally. The house had won awards and received lots of publicity and the firm seemed to be moving up a gear. Ferhan was busier than ever and expanding, yet it looked like we'd now be the only people ever to have an Azman Owens new-build. Joyce, though, had been divorced for several years and she was wearied by being a single parent so far from home. She had a huge, close family in Florida and they wanted her back.

I miss her, though. Every time the photograph on the piano gets pushed to the middle, I think about the architecture police.

Sometimes I nudge it along, as Joyce would have done. Sometimes, subversively, I leave it where it is.

The pain, like childbirth, receded. That summer, on holiday on the Mizen peninsula in south-west Ireland, we saw a remarkable amount of property for sale: down every other lane, some old barn or shack ripe for development.

'What I'd really like,' Charlie said idly one day as we were bowling along in the car to the beach, 'would be to build a house against a hillside, with views over the sea.'

I stared at him incredulously, this man who'd said never, never again: no more moving, not even any bathroom renovations.

'Glass at the front and slate behind . . .'

'I thought you never wanted to speak to a builder again.'

'Oh, I don't know,' Charlie said. 'I think we've got one more house left in us.' He slowed the car so he could look at me and smile. 'I think you think so too.'

Acknowledgements

With thanks to: Juliet Annan, Ferhan Azman, Elaine Bedell, David Bennett, Brian Eckersley, Georgia Garrett, Lux Patel, Joyce Owens and Meg Rosoff.

I am also indebted to writing by: Christopher Alexander, Gaston Bachelard, David Bennett, Stewart Brand, Clare Cooper Marcus, H. B. Cresswell, Carl Elliott, Suzanne Frank, Joel Garreau, Sarah Gaventa, Joyce Henri Robinson, Robert Hughes, Herbert and Katherine Jacobs, Philip Jodidio, Tracy Kidder, Henri Lefebvre, Kevin McCloud, Grant McCracken, Clare Melhuish, Michael Pollan, Christopher Reed, Witold Rybczynski, Richard Sennett, Catherine Slessor and Tom Wolfe.